SKILLS MASTERY & TEST PRACTICE

This Book Includes:

- **Access to Online SBAC Practice Assessments**
  - Two Performance Tasks (PT)
  - Two Computer Adaptive Tests (CAT)
  - Self-paced learning and personalized score reports
  - Strategies for building speed and accuracy
  - Instant feedback after completion of the Assessments
- **Standards based practice**
  - Ratios & Proportional Relationships
  - The Number System
  - Expressions & Equations
  - Geometry
  - Statistics & Probability
- **Detailed answer explanations for every question**

Complement Classroom Learning All Year

Using the Lumos Study Program, parents and teachers can reinforce the classroom learning experience for children. It creates a collaborative learning platform for students, teachers and parents.

Teacher Student

Parent

Used in Schools
To Improve Student Achievement

Lumos Learning

SBAC Test Prep: 7th Grade Math Common Core Practice Book and Full-length Online Assessments: Smarter Balanced Study Guide with Performance Task (PT) and Computer Adaptive Testing (CAT)

Contributing Author	-	Aaron Spencer
Contributing Author	-	Nikki McGee
Curriculum Director	-	Marisa Adams
Executive Producer	-	Mukunda Krishnaswamy
Designer	-	Mirona Jova
Database Administrator	-	R. Raghavendra Rao

ISBN-10: 1940484855

ISBN-13: 978-1-940484-85-3

Printed in the United States of America

For permissions and additional information contact us

Lumos Information Services, LLC
PO Box 1575, Piscataway, NJ 08855-1575
http://www.LumosLearning.com

Email: support@lumoslearning.com
Tel: (732) 384-0146
Fax: (866) 283-6471

Lumos Learning

Table of Contents

Introduction

The Common Core State Standards Initiative (CCSS) was created from the need to have more robust and rigorous guidelines which could be standardized from state to state. These guidelines create a learning environment where students will be able to graduate high school with all skills necessary to be active and successful members of society, whether they take a role in the workforce or in some sort of post-secondary education.

Once the CCSS were fully developed and implemented, it became necessary to devise a way to ensure they were assessed appropriately. To this end, states adopting the CCSS have joined one of two consortia, either PARCC or Smarter Balanced.

What is SBAC?

The Smarter Balanced Assessment Consortium (SBAC) is one of the two state consortiums responsible for developing assessments aligned to the rigorous Common Core State Standards. Thousands of educators, along with test developers, have worked together to create the new computer based English Language Arts and Math Assessments.

SBAC's first round of testing occurred during the 2014 – 2015 school year. The tests are conducted online, requiring students complete tasks to assess a deeper understanding of the CCSS and will involve a variety of new technology-enhanced question.

How Can the Lumos Study Program Prepare Students for SBAC Tests?

At Lumos Learning, we believe that year-long learning and adequate practice before the actual test are the keys to success on these standardized tests. We have designed the Lumos study program to help students get plenty of realistic practice before the test and to promote year-long collaborative learning.

This is a Lumos tedBook™. It connects you to Online SBAC Assessments and additional resources using a number of devices including Android phones, iPhones, tablets and personal computers. The Lumos StepUp Online Assessment is designed to promote year-long learning. It is a simple program students can securely access using a computer or device with internet access. Students will get instant feedback and can review their answers anytime. Each student's answers and progress can be reviewed by parents and educators to reinforce the learning experience.

How to access the Lumos SBAC Online Assessments

First Time Access:

Using a personal computer with internet access:	Using a smart phone or tablet:
Go to **http://www.lumoslearning.com/book** Enter the following access code in the Access Code field and press the Submit button. Access Code: SBACG7MATH-268-24-P	Scan the QR Code below and follow the instructions. 

Access Code: Please enter your Access Code [Submit]

In the next screen, click on the "New User" button to register your user name and password.

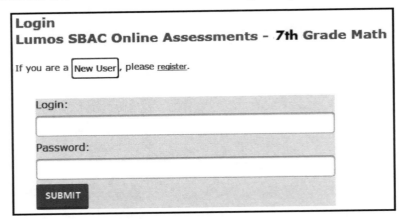

Subsequent Access:

After you establish your user id and password for subsequent access, simply login with your account information.

What if I buy more than one Lumos Study Program?

Please note that you can use all Online SBAC Assessments with one User ID and Password. If you buy more than one book, you will access them with the same account.

Go back to the **http://www.lumoslearning.com/book** link and enter the access code provided in the second book. In the next screen simply login using your previously created account.

How to create a teacher account

- You can use the Lumos online programs along with this book to complement and extend your classroom instruction.

- Get a Free Teacher account by visiting LumosLearning.com/a/sbacbasic

  This Lumos StepUp® Basic teacher account will help you:

  - Create up to 30 student accounts.
  - Review the online work of your students.
  - Easily access CCSS.
  - Create and share information about your classroom or school events.
  - Recommend useful mobile apps and books to your students.

  NOTE: There is a limit of one grade and subject per teacher for the free account.

- Download the Lumos SchoolUp™ mobile app using the instructions provided in "How can I Download the App?" section of this chapter.

- To learn more about the teacher portal please refer to "Lumos StepUp® Teacher Portal FAQ" section of this book.

QR code for Teacher account

Test Taking Tips

1) **The day before the test,** make sure you get a good night's sleep.

2) **On the day of the test,** be sure to eat a good hearty breakfast! Also, be sure to arrive at school on time.

3) **During the test:**

- **Read every question carefully.**

  - Do not spend too much time on any one question. Work steadily through all questions in the section.
  - Attempt all of the questions even if you are not sure of some answers.
  - If you run into a difficult question, eliminate as many choices as you can and then pick the best one from the remaining choices. Intelligent guessing will help you increase your score.
  - Also, mark the question so that if you have extra time, you can return to it after you reach the end of the section.
  - Some questions may refer to a graph, chart, or other kind of picture. Carefully review the graphic before answering the question.
  - Be sure to include explanations for your written responses and show all work.

- **While Answering Multiple-Choice (EBSR) questions.**

  - Select the bubble corresponding to your answer choice.
  - Read **all** of the answer choices, even if think you have found the correct answer.

- **While Answering TECR questions.**

  - Read the directions of each question. Some might ask you to drag something, others to select, and still others to highlight. Follow all instructions of the question (or questions if it is in multiple parts)

How to use this book effectively

The Lumos Program is a flexible learning tool. It can be adapted to suit a student's skill level and the time available to practice before standardized tests. Here are some tips to help you use this book and the online resources effectively:

Students

- The standards in each book can be practiced in the order designed, or in the order of your own choosing.
- Complete all problems in each workbook.
- Take the first practice CAT and PT online.
- Download the Lumos StepUp® app using the instructions provided in "How can I Download the App?" to have anywhere access to online resources.
- Have open-ended questions evaluated by a teacher or parent, keeping in mind the scoring rubrics.
- Take the second CAT and PT as you get close to the official test date.
- Complete the test in a quiet place, following the test guidelines. Practice tests provide you an opportunity to improve your test taking skills and to review topics included in the CCSS related standardized test.

Parents

- Familiarize yourself with the SBAC test format and expectations.
- Get useful information about your school by downloading the Lumos SchoolUp™ app. Please follow directions provided in "How can I Download the App?" section of this chapter.
- Help your child use Lumos StepUp® SBAC Online Assessments by following the instructions in "How to access the Lumos SBAC Online Assessments" section of this chapter.
- Review your child's performance in the "Lumos SBAC Online Assessments" periodically. You can do this by simply asking your child to log into the system online and selecting the subject area you wish to review.

Ratios and Proportional Relationships

Unit Rates (7.RP.A.1)

1. If y is proportional to x, and y = 4 when x = 6, what is the constant of proportionality between them (the ratio of x to y)?

   Ⓐ $\dfrac{4}{6}$

   Ⓑ $\dfrac{2}{3}$

   Ⓒ $\dfrac{3}{2}$

   Ⓓ 3

2. John eats a bowl of cereal for 3 of his 4 meals each day. He finishes two gallons of milk in eight days. How much milk does John use for one bowl of cereal? (Assume he only uses the milk for his cereal.)

   Ⓐ One-twelfth of a gallon of milk
   Ⓑ One cup of milk
   Ⓒ Two cups of milk
   Ⓓ One-sixth of a gallon of milk

3. A recipe to make a cake calls for three fourths of a cup of milk. Mary used this cake as the first layer of a wedding cake. The second layer was half the size of the first layer, and the third layer was half the size of the second layer. How much milk would be used for the entire wedding cake?

   Ⓐ One and two-thirds cups of milk
   Ⓑ One and one-third cups of milk
   Ⓒ One and five-sixteenths cups of milk
   Ⓓ One cup of milk

4. **One third of a quart of paint covers one fourth of a basketball court. How much paint does it take to paint the entire basketball court?**

    Ⓐ one and one-third quarts
    Ⓑ one quart
    Ⓒ one and one-fourth quarts
    Ⓓ one and three-fourths quarts

5. **The total cost of 100 pencils purchased at a constant rate is $39.00. What is the unit price?**

    Ⓐ $39.00
    Ⓑ $3.90
    Ⓒ $0.39
    Ⓓ $0.039

6. **A construction worker was covering the bathroom wall with tiles. He covered three-fifths of the wall with 50 tiles. How many tiles will it take to cover the entire wall?**

    Ⓐ 83 tiles
    Ⓑ 83 and one-third tiles
    Ⓒ 85 tiles
    Ⓓ 83 and one-half tiles

7. **Jim ran four-fifths of a mile and dropped out of the 1600 meter race. His pace was 12 miles an hour until the point he dropped out of the race. How many minutes did he run?**

    Ⓐ 4 minutes and 30 seconds
    Ⓑ 4 minutes
    Ⓒ 4 minutes and 20 seconds
    Ⓓ 4 minutes and 10 seconds

8. **Ping played three-fourths of a football game. The game was three and a half hours long. How many hours did Ping play in this game?**

    Ⓐ 2 hours 37 minutes
    Ⓑ 2 hours 37 minutes and 30 seconds
    Ⓒ 2 hours 37 minutes and 20 seconds
    Ⓓ 2 hours 37 minutes and 10 seconds

9. Bill is working out by running up and down steps at the local stadium. He runs a different number of steps in a random order.

    Which of the following is his best time?

    Ⓐ 25 steps in 5 minutes
    Ⓑ 30 steps in 5.5 minutes
    Ⓒ 20 steps in 4.5 minutes
    Ⓓ 15 steps in 4 minutes

10. Doogle drove 30 and one-third miles toward his brother's house in one-third of an hour. About how long will the entire 100-mile trip take at this constant speed?

    Ⓐ 1 hour
    Ⓑ 1 hour and 6minutes
    Ⓒ 1 hour and 1minutes
    Ⓓ 1 hour and 3 minutes

Understanding and Representing Proportions

1. The following table shows two variables in a proportional relationship:

a	b
2	6
3	9
4	12

Which of the following is an algebraic statement showing the relationship between a and b.

Ⓐ a = 3b
Ⓑ b = 3a
Ⓒ b = 1/3 (a)
Ⓓ a = 1/2 (b)

2. If the ratio of the length of a rectangle to its width is 3 to 2, what is the length of a rectangle whose width is 4 inches?

Ⓐ 4 in.
Ⓑ 5 in.
Ⓒ 6 in.
Ⓓ 7 in.

3. The following table shows two variables in a proportional relationship:

e	f
5	25
6	30
7	35

Using the relationship between e and f as shown in this table, find the value of f when e = 11.

Ⓐ 40
Ⓑ 45
Ⓒ 50
Ⓓ 55

4. If a vehicle is traveling at a constant rate, and it takes 1.5 hours to travel 97.5 miles, how long does it take to travel 130 miles?

   Ⓐ 2 hours
   Ⓑ 2.5 hours
   Ⓒ 3 hours
   Ⓓ 3.5 hours

5. Ricky's family wants to invite his classroom to a "get acquainted" party. If 20 students attend, the party will cost $100. Assuming the relationship between cost and guests is proportional, which of the following will be the cost if 29 students attend?

   Ⓐ $129
   Ⓑ $135
   Ⓒ $139
   Ⓓ $145

6. Which of the following pairs of ratios form a proportion?

   Ⓐ 9 boys to 5 girls and 12 boys to 8 girls
   Ⓑ 9 boys to 5 girls and 18 boys to 10 girls
   Ⓒ 9 boys to 5 girls and 13 boys to 9 girls
   Ⓓ 9 boys to 5 girls and 27 boys to 14 girls

7. If the local supermarket is selling oranges for *p*, cents each and Mrs. Jones buys *n*, oranges, write an equation for the total *(t)*, that Mrs. Jones pays for oranges.

   Ⓐ $t = p + n$
   Ⓑ $t = p - n$
   Ⓒ $t = pn$
   Ⓓ $t = p/n$

8. The ratio of Kathy's earnings to her hours worked is a constant. This fact implies which of the following?

   Ⓐ Kathy's earnings are proportional to her hours of work.
   Ⓑ If Kathy works harder during her 8 hours of work today, her income for today will be greater than if she just takes it easy.
   Ⓒ If Kathy only works a half day today, her earnings will be the same as if she worked all day.
   Ⓓ If Kathy takes an extra hour for lunch, it will not affect her earnings for the day.

9. **Write a mathematical statement (equation) for the relationship between feet and yards.**

    Ⓐ number of feet / number of yards = 3
    Ⓑ number of yards / number of feet = 3
    Ⓒ number of feet - number of yards = 3
    Ⓓ number of feet + number of yards = 3

10. **The ratio of the measurement of a length in yards to the measurement of the same length in feet is a constant. This implies that __?__**

    Ⓐ the measurement of a length in yards is directly proportional to the measurement of the same length in feet.
    Ⓑ the measurement in yards of a length is inversely proportional to the measurement of the same length in feet.
    Ⓒ the measurement in yards of a length is equivalent to the measurement of the same length in feet.
    Ⓓ There is no relationship between the two measurements.

Finding Constant of Proportionality (7.RP.A.2.A)

1. According to the graph, what is the constant of proportionality?

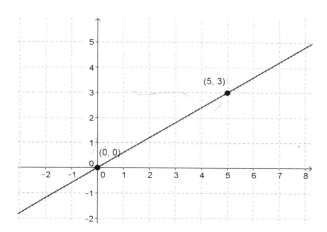

(A) $\dfrac{3}{5}$

(B) 5

(C) 3

(D) $\dfrac{1}{3}$

2. According to the table, how much does one ticket cost?

Number of Tickets	Total Cost
3	$ 21.00
4	$ 28.00
5	$ 35.00
6	$ 42.00

(A) $21.00

(B) $4.75

(C) $7.00

(D) $16.50

3. **What is the unit rate for a pound of seed?**

Pounds of Seed	Total Cost
10	$ 17.50
20	$ 35.00
30	$ 52.50
40	$ 70.00

Ⓐ $3.50
Ⓑ $1.75
Ⓒ $17.50
Ⓓ $7.25

4. **If y = 3x, what is the constant of proportionality between y and x?**

Ⓐ 1
Ⓑ 0.30
Ⓒ 1.50
Ⓓ 3

5. **When Frank buys three packs of pens, he knows he has 36 pens. When he buys five packs, he knows he has 60 pens. What is the constant of proportionality between the number of packs and the number of pens?**

Ⓐ 12
Ⓑ 10
Ⓒ 36
Ⓓ 60

6. **According to the table, how many eggs are collected per month from one chicken?**

Number of Chickens	Total Eggs Per Month
8	160
10	200
12	240
14	280

Ⓐ 200
Ⓑ 160
Ⓒ 16
Ⓓ 20

7. **What is the unit rate for the number of hours of study each week per class?**

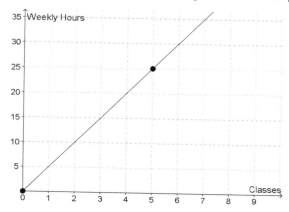

- Ⓐ 25 hours
- Ⓑ 5 hours
- Ⓒ 7 hours
- Ⓓ 10 hours

8. **What is the constant of proportionality in the following equation?**

    B = 1.25C

- Ⓐ 1.00
- Ⓑ 0.25
- Ⓒ 1.25
- Ⓓ 2.50

9. **What is the unit rate for the cups of flour per recipe?**

Number of Recipes	Cups of Flour
1.5	6
2	8
2.5	10
3	12

- Ⓐ 6 cups
- Ⓑ 2 cups
- Ⓒ 8 cups
- Ⓓ 4 cups

10. When Georgia buys 3 boxes of peaches, she has 40 pounds more than when she buys 1 box of peaches. How many pounds of peaches are in each box?

Ⓐ 40
Ⓑ 60
Ⓒ 20
Ⓓ 3

Represent Proportions by Equations (7.RP.A.2.B)

1. **3 hats cost a total of $18. Which equation describes the total cost, C, in terms of the number of hats, n?**

   Ⓐ C = 3n
   Ⓑ C = 6n
   Ⓒ C = 0.5n
   Ⓓ 3C = n

2. **Use the data in the table to give an equation to represent the proportional relationship.**

x	y
0.5	7
1	14
1.5	21
2	28

   Ⓐ y = 14x
   Ⓑ y = 7x
   Ⓒ 7y = x
   Ⓓ 21y = x

3. **Kelli has purchased a membership at the gym for the last four months. She has paid the same amount each month, and her total cost so far has been $100. What equation expresses the proportional relationship of the cost per month?**

   Ⓐ C = 100m
   Ⓑ C = 50m
   Ⓒ C = 4m
   Ⓓ C = 25m

4. **When buying bananas at the market, Marco pays $4.50 for 5 pounds. What is the relationship between pounds, p, and cost, C.**

   Ⓐ C = 4.5p
   Ⓑ C = 5p
   Ⓒ C = 0.9p
   Ⓓ C = 22.5p

5. The cost to rent an apartment is proportional to the number of square feet in the apartment. An 800 square foot apartment costs $600 per month. What equation represents the relationship between area, a, and cost, C?

    Ⓐ C = 0.75a
    Ⓑ C = 1.33a
    Ⓒ C = 8a
    Ⓓ C = 6a

6. A school has to purchase new desks for their classrooms. They have to purchase 350 new desks, and they pay $7000. What equation demonstrates the relationship between the number of desks, d, and the total cost, C?

    Ⓐ C = 10d
    Ⓑ C = 20d
    Ⓒ C = 70d
    Ⓓ C = 35d

7. A soccer club is hosting a tournament with 12 teams involved. Each team has a set number of players, and there are a total of 180 players involved in the tournament. Which equation represents the proportional relationship between teams, t, and players, p?

    Ⓐ p = 15t
    Ⓑ p = 12t
    Ⓒ p = 18t
    Ⓓ p = 11t

8. A pack of three rolls of tape costs $5. What is the proportional relationship between cost, C, and packs of tape, p?

    Ⓐ C = 1.67p
    Ⓑ C = 15p
    Ⓒ C = 3p
    Ⓓ C = 5p

9. Freddy is building a house and has 5 loads of gravel delivered for the work. His total cost for the gravel is $1750. Which equation correctly shows the relationship between cost, C, and loads of gravel, g?

    Ⓐ C = 5g
    Ⓑ C = 350g
    Ⓒ C = 75g
    Ⓓ C = 225g

10. Use the graph to write an equation for the proportional relationship between the number of hours Corrie works, h, and her total pay, P.

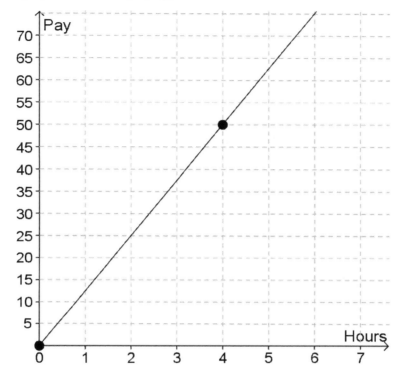

Ⓐ   P = 12.50h
Ⓑ   P = 50h
Ⓒ   P = 4h
Ⓓ   P = 22.50h

Significance of Points on Graphs of Proportions (7.RP.A.2.D)

1. **Which point on the graph of the straight line demonstrates that the line represents a proportion?**

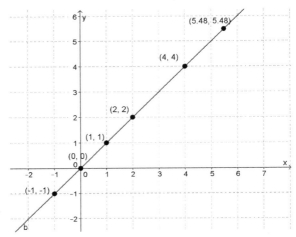

    Ⓐ (2, 2)
    Ⓑ (0, 0)
    Ⓒ (5.48, 5.48)
    Ⓓ (-1, -1)

2. **Which point on the graph of the straight line names the unit rate of the proportional relationship?**

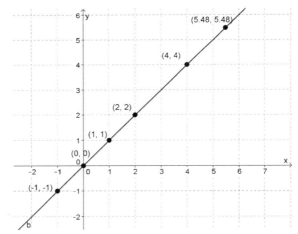

    Ⓐ (1, 1)
    Ⓑ (0, 0)
    Ⓒ (2, 2)
    Ⓓ (4, 4)

3. **The graph shows the relationship between the number of classes in the school and the total number of students. How many students are there per class?**

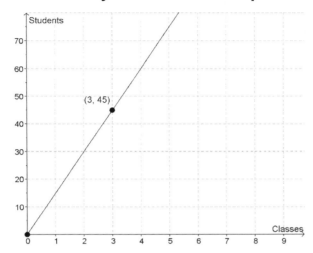

Ⓐ 45 students
Ⓑ 3 students
Ⓒ 135 students
Ⓓ 15 students

4. **Use the information given on the graph to find the value of y.**

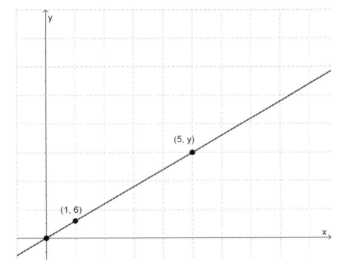

Ⓐ 30
Ⓑ 5
Ⓒ 10
Ⓓ 11

5. **In order for the relationship to be proportional, what other point must be a part of the graph?**

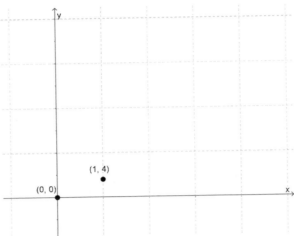

- Ⓐ (5, 10)
- Ⓑ (5, 25)
- Ⓒ (5, 15)
- Ⓓ (5, 20)

6. **What is the unit rate of Birthday presents per child?**

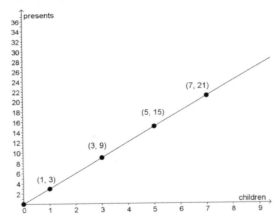

- Ⓐ 21 presents
- Ⓑ 9 presents
- Ⓒ 3 presents
- Ⓓ 15 presents

7. **What is the unit rate of bushels of apples per tree?**

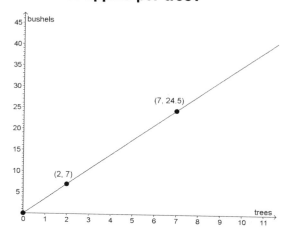

- Ⓐ 3.5
- Ⓑ 7
- Ⓒ 24.5
- Ⓓ 14

8. **Which point represents the profit if no boxes of popcorn are sold for the fundraiser?**

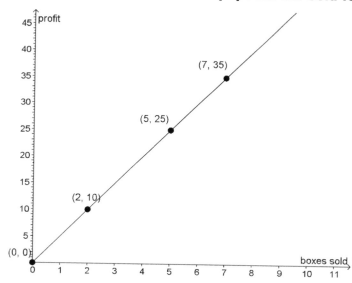

- Ⓐ (5, 25)
- Ⓑ (7, 35)
- Ⓒ (0, 0)
- Ⓓ (2, 10)

9. **If the relationship between x and y is proportional, what point on the line will indicate the unit rate of that relationship?**

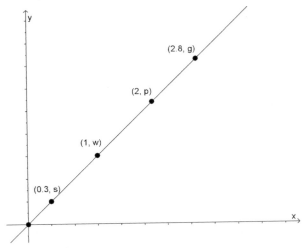

   Ⓐ (2, p)
   Ⓑ (0.3, s)
   Ⓒ (2.8, g)
   Ⓓ (1, w)

10. **According to the information given on the graph, how much would 20 boards cost?**

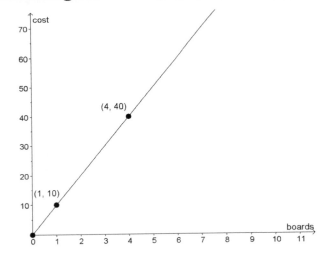

   Ⓐ $200
   Ⓑ $400
   Ⓒ $80
   Ⓓ $120

Applying Ratios and Percents (7.RP.A.3)

1. **What value of *x* will make these two expressions equivalent?**

$$\frac{-3}{7} \text{ and } \frac{x}{21}$$

Ⓐ x = -3
Ⓑ x = 7
Ⓒ x = 9
Ⓓ x = -9

2. **If *p* varies proportionally to *s*, and *p* = 10 when *s* = 2, which of the following equations correctly models this relationship?**

Ⓐ p = 5s
Ⓑ p = 10s
Ⓒ s = 10p
Ⓓ 2s = 10p

3. **Solve for x, if** $\dfrac{72}{108}$ **and** $\dfrac{x}{54}$ **are equivalent.**

Ⓐ x = 18
Ⓑ x = 36
Ⓒ x = 54
Ⓓ x = 24

4. **At one particular store, the sale price, *s*, is always 75% of the displayed price, *d*. Which of the following equations correctly shows how to calculate *s* from *d*?**

Ⓐ d = 75s
Ⓑ s = 0.75d
Ⓒ s = d - 0.75
Ⓓ s = d + 75

5. **When *x* = 6, *y* = 4. If *y* is proportional to *x*, what is the value for *y* when *x* = 9?**

Ⓐ 4

Ⓑ $\dfrac{2}{3}$

Ⓒ 3

Ⓓ 6

6. Jim is shopping for a suit to wear to his friend's wedding. He finds the perfect one on sale at 30% off. If the original price was $250.00, what will the selling price be after the discount?

    Ⓐ $75
    Ⓑ $175
    Ⓒ $200
    Ⓓ $220

7. If Julie bought her prom dress on sale at 15% off and paid $110.49 before tax, find the original price of her dress.

    Ⓐ $126.55
    Ⓑ $129.99
    Ⓒ $135.00
    Ⓓ $139.99

8. A plot of land is listed for sale with the following measurements: 1300 ft x 982 ft. When the buyer measured the land, he found that it measured 1285 ft by 982 ft. What was the % of error of the area of the plot?

    Ⓐ 1.47%
    Ⓑ 14.73%
    Ⓒ 1.17%
    Ⓓ 11.7%

9. Pierre received a parking ticket with an original fee of $22.00. Each month that he failed to make payment, fees of $7.00 were added. By the time he paid the ticket, his bill was $36.00. What was the ratio of fees to the cost of the ticket?

    Ⓐ 36/22
    Ⓑ 22/36
    Ⓒ 7/22
    Ⓓ 7/11

10. Sara owns a used furniture store. She bought a chest for $42 and sold it for $73.50. What percent did she mark up the chest?

    Ⓐ 100%
    Ⓑ 75%
    Ⓒ 42%
    Ⓓ 31.5%

End of Ratios and Proportional Relationships

Ratios and Proportional Relationships

Answer Key
&
Detailed Explanations

Unit Rates (7.RP.A.1)

Question No.	Answer	Detailed Explanation
1	C	3/2 is the constant of proportionality when y = 4 and x = 6. To find the constant of proportionality, apply the concept of unit rate. Since y is proportional to x, that means for every 6 of x, there is 4 of y. Simplifying the ratio 6/4 = 3/2.
2	A	First, find out how much cereal John eats in 8 days: (1) 3 bowls per day x 8 days = 24 bowls. Since it takes 2 gallons of milk to eat 24 bowls of cereal, set up the ratio and simplify: (2) 2/24 (GCF is 2, so divide numerator and denominator to find simplest form) (3) 1/12 Therefore, John uses 1/12 of a gallon of milk in each bowl of cereal.
3	C	In order to solve this problem, multiply 3/4 × 1/2 and then multiply the product times 1/2 in order to find the amount of milk used for each section of the cake. (1) 3/4 × 1/2 = 3/8 (2) 3/8 × 1/2 = 3/16 The amount of milk used is 3/4 + 3/8 + 3/16. (3) 3/4 + 3/8 + 3/16 -- The LCD is 16, so rename 3/4 (multiply both numerator and denominator by 4) and 3/8 (multiply both numerator and denominator by 2). (4) 12/16 + 6/16 + 3/16 = (5) 18/16 + 3/16 = (6) 21/16 (convert the improper fraction into a mixed number) (7) 1 5/16 Therefore, one and five-sixteenths cups of milk would be used to make the cake.
4	A	To solve this problem, add 1/3 + 1. Use 1 because if 1/3 covers 1/4 of the court, then 1/3 + 1 will cover the entire court (1) 1/3 + 1/1 The LCD is 3, so rename 1/1 (multiply both numerator and denominator by 3) (2) 1/3 + 3/3 = (3) (1 + 3) / 3 (4) 4/3 --- convert into a mixed number (5) 1 1/3 Therefore, it will take one and one-third quarts to cover the entire basketball court.
5	C	Unit price means price per one unit. Therefore, we need to know the price per pencil. Since the price of the pencils is defined to be a constant rate, then the total cost ($39.00) divided by the total number of pencils (100) will give us the cost per pencil (or per unit). $39.00 / 100 = $0.39 per pencil The unit price is $0.39. The correct answer is $0.39.

Question No.	Answer	Detailed Explanation
6	B	To solve this problem, find out how much it takes to cover 1/5 of the wall, and then multiply by 5. (1) 50 ÷ 3 = (how much it takes to cover 1/5 of the roof) (2) multiply 16.66 × 5 = 83.33 (3) 83.33 = 83 1/3 Therefore, it will take 83 and one-third tiles to cover the entire wall.
7	B	To solve this problem, set up a proportion: (distance)/(time) = (distance)/(time) Plug in the numbers (you can convert to decimals to simplify the process): (0.8 mile)/(x min) = (12 miles)/(60 min). Using cross products a/b = c/d is ad = bc (0.8)(60) = (x)(12) 48 = 12x (solve for x by dividing each side by 12)4 = x. Therefore, Jim ran 4 minutes.
8	B	To solve this problem, multiply 3/4 × 3 1/2(1). Convert 3 1/2 to an improper fraction = 7/2(2) 3/4 × 7/2 =(3) (3 × 7)/(4 × 2) =(4) 21/8 (convert back to a mixed number)(5) 2 5/8(6) 5/8 = 37.5 minutesTherefore, Ping played 2 hours 37 minutes and 30 seconds.
9	B	50/5 = 10 steps per minute 60/5.5 = 10.9 steps per minute * 40/4.5 = 8.9 steps per minute 30/4 = 7.5 steps per minute 30 steps in 5.5 minutes gives an average of 10.9 steps per minute, which is his best time. 30 steps in 5.5 minutes is the correct answer.
10	B	To solve this problem, set up a proportion: (distance)/(time) = (distance)/(time) 1/3 of a hour is equivalent to 20 minutes. Plug in the numbers (you can convert to decimals to simplify the process): (30.33 miles)/(20 minutes) = (100 miles)/(x minutes). Using cross products a/b = c/d is ad = bc (30.33)(x) = (20)(100) 30.33x = 2000 (solve for x by dividing each side by 30.33) x = 65.94 (round up to the nearest whole number). Since 65.94 is almost 66 minutes, that is the same as one hour and 6 minutes.

Understanding and Representing Proportions (7.RP.A.2.A)

1	B	In each case, if we multiply a by 3, we get b. b = 3a is the correct answer.
2	C	length/width = 3 / 2 =x / 4 2x = 12 x = 6 in. 6 in. is the correct answer.
3	D	Since the table shows that f = 5e, then 5 is the constant of proportionality. Therefore, f = 5e; so if e = 11, f = 55. 55 is the correct answer.
4	A	Since rate is constant, d/t is constant. 97.5/1.5 = 130/t 97.5t = (130)(1.5) t = 195/97.5 t = 2 hours. 2 hours is the correct answer.

Question No.	Answer	Detailed Explanation
5	D	Since the relationship is proportional we can write the mathematical statement, 20/100 = 29/c. So 20c = 2900. c = $145. $145 is the correct answer.
6	B	A proportion consists of two equivalent ratios. 9/5 = 18/10 because 9 x 10 = 5 x 18. 9 boys to 5 girls and 18 boys to 10 girls is the correct answer.
7	C	In this case, p is constant regardless of the number, n, of oranges that she buys. To find the total cost, t, we multiply the number of oranges purchased, n, by the cost per orange, p. t = pn is the correct answer.
8	A	If Kathy's earnings are proportional to her hours of work, then the ratio of her earnings to hours worked is a constant. "Kathy's earnings are proportional to her hours of work" is the correct answer.
9	A	Since 3 ft = 1 yd, if we divide feet by yards, the quotient will always be 3. "number of feet / number of yards = 3" is the correct answer.
10	A	Since these ratios are always equivalent to a constant, the variables are directly proportional by definition. "The measurement in yards of a length is directly proportional to the measurement of the same length in feet" is the correct answer.

Finding Constant of Proportionality (7.RP.A.2.B)

1	A	The line passes through the origin, so it is a proportional relationship. The second point gives the constant of proportionality, which is the y value divided by the x value: 3/5.
2	C	For any of the pairs of data, divide the total cost by the number of tickets to find how much one ticket costs. For example, $21/3 = $7 per ticket.
3	B	For any of the pairs of data, divide the total cost by the number of pounds of seed to find how much one pound costs. For example, $17.50/10 = $1.75 per pound.
4	D	Because the equation is solved for y and is equal to a multiple of x, the coeffecient of the x term gives the constant of proportionality: 3.
5	A	Divide the number of pens by the number of packs for either pair of data. For example, 36/3 = 12 pens per pack.

Question No.	Answer	Detailed Explanation
6	D	Divide the total eggs per month by the number of chickens for any of the pairs of data. For example, 160/8 = 20 eggs per chicken per month.
7	B	The line passes through the origin, so it is a proportional relationship. The second point gives the constant of proportionality, which is the y value divided by the x value: 25/5, which simplifies to 5.
8	C	Because the equation is solved for y and is equal to a multiple of x, the coeffecient of the x term gives the constant of proportionality: 1.25.
9	D	Don't simply look at the increase from one entry to the next because the number of recipes does not increase in whole units. Rather, divide the cups of flour by the number of recipes. For example, 8/2 = 4 cups of flour per recipe.
10	C	The difference between our two pairs of data is 2 boxes and 40 pounds of peaches. Divide the difference in pounds by the difference in boxes: 40/2 = 20 pounds per box.

Represent proportions by equations (7.RP.A.2.C)

1	B	Divide the total cost by the number of hats: 18/3 = 6. This gives the cost per hat, which is the constant of proportionality in the equation: C = 6n.
2	A	Choose a pair of data. Divide the value of y by the value of x: 14/1 = 14. This gives the constant of proportionality in the equation: y = 14x.
3	D	Divide the total cost by the number of months: 100/4 = 25. This gives the cost per month, which is the constant of proportionality in the equation: C = 25m.
4	C	Divide the total cost by the number of pounds: 4.50/5 = 0.9. This gives the cost per pound, which is the constant of proportionality in the equation: C = 0.9p.
5	A	Divide the rental cost by the area of the apartment: 600/800 = 0.75. This gives the cost per square foot of area, which is the constant of proportionality in the equation: C = 0.75a.
6	B	Divide the total cost by the number of desks: 7000/350 = 20. This gives the cost per desk, which is the constant of proportionality in the equation: C = 20d.

Name: _____   Date: _____

Question No.	Answer	Detailed Explanation
7	A	Divide the total number of players by the number of teams: 180/12 = 15. This gives the number of players per team, which is the constant of proportionality in the equation: p = 15t.
8	D	The relationship named is that between cost and the number of packs. The number of rolls per pack is irrelevant. The cost per pack is already given as $5 per pack, so this is the constant of proportionality in the equation: C = 5p
9	B	Divide the total cost by the number of loads: 1750/5 = 350. This gives the cost per load, which is the constant of proportionality in the equation: C = 350g.
10	A	The straight line passes through the origin, so the relationship is proportional. Use the values of a point on the line. Divide the pay by the number of hours: 50/4 = 12.50. This gives the pay per hour, which is the constant of proportionality in the equation: P = 12.50h.

Significance of points on graphs of proportions (7.RP.A.2.D)

1	B	A straight line graph is a proportion if and only if it passes through the origin, (0, 0).
2	A	The unit rate is the amount per single unit of something. This means that the point located at 1 on the x-axis will indicate the unit rate: (1, 1).
3	D	Find the unit rate, which is the number of students per one class, by dividing the y value by the x value for the given point: 45/3 = 15 students per class.
4	A	The unit rate shows that y is 6 times the value of x. Multiply the x value of 5 by 6, and 30 is the value of y.
5	D	The unit rate is given by the point (1, 4). y must be 4 times x. That relationship is found only in the point (5, 20).
6	C	The unit rate is found at the point where x has a value of 1. That point is (1, 3), so the unit rate is 3 presents per child.
7	A	The unit rate can be found by dividing the y value by the x value for any given point: 7/2 = 3.5 bushels per tree.
8	C	The point (0, 0) always represents the origin of the proportion. In this case, that means that no boxes are sold, and the profit will be $0.

32   © Lumos Information Services 2015 | LumosLearning.com

Question No.	Answer	Detailed Explanation
9	D	The unit rate is always located at the point where the x value is 1. In this case, that is (1, w).
10	A	The unit rate is found at (1, 10), which indicates that 1 board has a cost of $10. Multiply by 20 boards to find the total cost: (10)(20) = $200.

Applying Ratios and Percents (7.RP.A.3)

Question No.	Answer	Detailed Explanation
1	D	In a set of equivalent ratios, or a proportion, the numerator and denominator of one ratio will be multipled by the same number to get the values of the other ratio. In this case, the denominator of the first ratio, 7, is multiplied by 3 to get to 21. This means (-3) must also be multiplied by 3 to get to (-9).
2	A	To find the constant of proportionality, find the relationship between p and s. When p = 10 and s = 2, dividing p by s shows that p is 5 times s. Therefore, the equation that shows the constant of proportionality is p = 5s.
3	B	To solve this proportion for x, multiply both sides of the equation by 54, and simplify the result.
4	B	To find the constant of proportionality, find the relationship between s and d. s is 75% of d, which is the same as 0.75 times d. Therefore, the equation that shows the constant of proportionality is s = 0.75d.
5	D	Set up a proportion between the known ratio and the unknown ratio, and solve for y. $6/4 = 9/y$ $6y = 9\,(4)$ Cross Multiply $6y = 36$ Simplify $6y/y = 36/6$ Divide each side by 6 $y = 6$ Simplify
6	B	Selling price (sp) = Original price (op) - 0.30(op) $sp = 250 - 0.30(250)$ $sp = 250 - 75$ $sp = 175$. $175 is the correct answer.
7	B	Original price (op) - 0.15(op) = Selling price (sp) $op - 0.15(op) = 110.49$ $0.85(op) = 110.49$ $op = 110.49 / 0.85$ $op = \$129.99$. $129.99 is the correct answer.

Question No.	Answer	Detailed Explanation
8	C	Using the listed measurements: Area = 1300 x 982 = 1,276,600 square feet. Using the actual measurements: Area = 1285 x 982 = 1,261,870 square feet. Error = 14,730 square feet. To find the percent error, divide the amount of error by the listed area: 14,730 / 1,276,600 = 1.15% error. The correct answer is 1.15%.
9	D	$36.00 - $22.00 = $14.00 in fees. $14.00 / $22.00 = 14 / 22 = 7 /11 is the ratio of fees to the cost of the ticket. 7 / 11 is the correct answer.
10	B	Selling price - cost = markup $73.50 - $42.00 = $31.50 markup markup / (cost - percent of markup) $31.50 / $42 = 0.75 = 75% markup. The correct answer is 75%.

The Number System

1. **Which picture represents the rational number** $\dfrac{1}{2}$ **?**

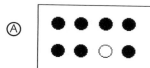

Ⓐ

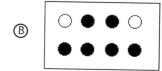

Ⓑ

Ⓒ

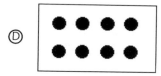

Ⓓ

2. **Evaluate: 25 + 2.005 - 7.253 - 2.977**

Ⓐ -16.775
Ⓑ 16.775
Ⓒ 167.75
Ⓓ 1.6775

3. Add and/or subtract as indicated : $-3\dfrac{4}{5} + 9\dfrac{7}{10} - 2\dfrac{11}{20} =$

Ⓐ $3\dfrac{7}{20}$

Ⓑ $4\dfrac{7}{10}$

Ⓒ $4\dfrac{9}{20}$

Ⓓ $3\dfrac{1}{20}$

4. Find the difference of 268.75 and - 46.99.

Ⓐ 221.76
Ⓑ 222.24
Ⓒ 315.74
Ⓓ 314.74

5. If p and q represent different numbers, which of the following expressions combine to equal q?

Ⓐ p - (p - q)
Ⓑ p - (q - p)
Ⓒ q - (q - p)
Ⓓ q - (p - q)

6. Linda and Carrie made a trip from their hometown to a city about 200 miles away to attend a friend's wedding. The following chart shows their distances, stops and times. What part of their total trip did they spend driving?

3hr	driving
15 min	rest stop
1 1/2 hr	driving
1 hr	res stop
20 min	driving

Ⓐ 4/5
Ⓑ 2/5
Ⓒ 58/73
Ⓓ 99/100

7. If $a = \dfrac{5}{6}$, $b = -\dfrac{2}{3}$ and $c = -1\dfrac{1}{3}$, find a - b - c.

Ⓐ - 1 1/6
Ⓑ 2 5/6
Ⓒ - 2 1/6
Ⓓ - 2 5/6

8. Jim is fencing a rectangular yard 36 ft by 28 ft. He has 150 ft of fencing and will be installing a 5 ft gate on one end. How much fencing will he have left?

Ⓐ 86 ft
Ⓑ 91 ft
Ⓒ 123 ft
Ⓓ 27 ft

9. If Ralph ate half of his candy bar followed by half of the remainder followed by half of that remainder, what part was left?

Ⓐ 1/4
Ⓑ 1/8
Ⓒ 1/6
Ⓓ 1/3

10. Mary is making a birthday cake to surprise her mom. She needs $3\dfrac{1}{2}$ cups of flour but she only has $\dfrac{1}{3}$ cup. How much more flour does she need?

Ⓐ $3\dfrac{1}{3}$ cups

Ⓑ $3\dfrac{1}{4}$ cups

Ⓒ $3\dfrac{1}{6}$ cups

Ⓓ 3 cups

Add and Subtract Rational Numbers (7.NS.A.1.B)

1. If p + q has a value that is exactly 1/3 less than p, what is the value of q?

   Ⓐ –1/3
   Ⓑ 2/5
   Ⓒ 1/3
   Ⓓ –2/5

2. What is the sum of k and the opposite of k?

   Ⓐ 2k
   Ⓑ k + 1
   Ⓒ 0
   Ⓓ –1

3. If p + q has a value of 12/5, and p has a value of 4/5, what is the value of q?

   Ⓐ 5/8
   Ⓑ 8/5
   Ⓒ 1/3
   Ⓓ 3/2

4. What is the value of b in the diagram?

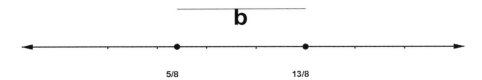

   Ⓐ 0.8
   Ⓑ 1
   Ⓒ 1.8
   Ⓓ 2.2

5. t has a value of 5/2. p is the sum of t and v, and p has a value of 0. What is the value of v?

   Ⓐ –1/3
   Ⓑ 4
   Ⓒ 2.5
   Ⓓ –5/2

6. **What is the value of n in the diagram?**

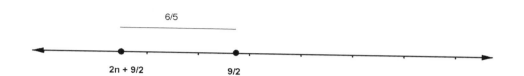

Ⓐ $\dfrac{6}{5}$

Ⓑ 0.6

Ⓒ $\dfrac{-3}{5}$

Ⓓ 2.1

7. **What is the sum of 10 and –10?**

Ⓐ 20

Ⓑ 0

Ⓒ 1

Ⓓ –20

8. **If the value of r is –1/3 and t has a value of –1/2, will t + r be to the right or left of t on a number line, and why?**

Ⓐ To the left because the absolute value of r is less than the absolute value of t.

Ⓑ To the right because the absolute value of r is less than the absolute value of t, and both numbers are negative.

Ⓒ To the left because r is a negative value being added to t.

Ⓓ To the left because the sum of two negative numbers is always negative.

9. **If the value of A is 3, and the value of B is –2/3, how far apart will A and A + B be on a number line?**

Ⓐ 7/3

Ⓑ 2/3

Ⓒ 5/3

Ⓓ 11/3

10. **The distance between H and K on a number line is 9/4. If K has a value of 7/4, which of the following might be the value of H?**

Ⓐ 2/4

Ⓑ 9/4

Ⓒ –4

Ⓓ –1/2

Additive Inverse and Distance Between Two Points on a Number Line (7.NS.A.1.C)

1. **Which of the following is the same as 7 – (3 + 4)?**

    Ⓐ 7 + (−3) + (−4)
    Ⓑ 7 +(−3 + 4)
    Ⓒ 7 + 7
    Ⓓ −7 – 7

2. **Which of the following expressions represents the distance between the two points?**

    Ⓐ | 4 - 3 |
    Ⓑ (-3) -4
    Ⓒ | (-3) - 4 |
    Ⓓ 4 - 3

3. **Kyle and Mark started at the same location. Kyle traveled 5 miles due east, while Mark traveled 3 miles due West. How far apart are they?**

    Ⓐ 2 miles
    Ⓑ 8 miles
    Ⓒ 15 miles
    Ⓓ 12 miles

4. **The distance between G and H on the number line is | 5 - (- 2) |. What might be the coordinate of H?**

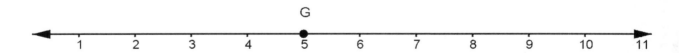

    Ⓐ 2
    Ⓑ 7
    Ⓒ 5
    Ⓓ −2

5. **Which of the following is the same as 2x – 3y – z?**

    Ⓐ 2x – (3y – z)
    Ⓑ 2x + (–3y) – z
    Ⓒ 2x + 3y – z
    Ⓓ (–2x) – 3y – z

6. **If Brad lives 5 blocks north of the park and Easton lives 7 blocks south of the park, which of the following correctly represents an expression for the distance between their homes?**

    Ⓐ | 5 - (-7) |
    Ⓑ 7 - 5
    Ⓒ 3 - (- 4)
    Ⓓ | 5 + (- 7) |

7. **For what numbers will the statement be true?: t – (w) = t + (–w)**

    Ⓐ All positive numbers
    Ⓑ All positive integers
    Ⓒ All positive and negative integers, but not 0
    Ⓓ All real numbers

8. **Which of the following is the same as 4 + (x – 3)?**

    Ⓐ 4 + (–x) – 3
    Ⓑ (–4) + x + (–3)
    Ⓒ 4 + x + (–3)
    Ⓓ 1 + (–x)

9. **Which of the following expressions represents the distance between the two points?**

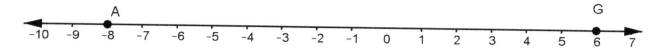

    Ⓐ |6-8|
    Ⓑ (- 8)+6
    Ⓒ (- 8)- 6
    Ⓓ |6 - (-8)|

10. Which words best complete the statement?

The distance between two numbers on a number line is the same as the _____ of their _____.

Ⓐ Absolute value; difference
Ⓑ Sum; squares
Ⓒ Difference; squares
Ⓓ Absolute value; sum

Strategies for Adding and Subtracting Rational Numbers (7.NS.A.1.D)

1. **What property is illustrated in the equation?:**

$$7 + \frac{1}{2} + \frac{1}{2} = \frac{1}{2} + \frac{1}{2} + 7$$

  Ⓐ Commutative property of addition
  Ⓑ Associate property of addition
  Ⓒ Distributive property
  Ⓓ Identity property of addition

2. **Which is a valid use of properties to make the expression easier to calculate?**

  $92 - 8 = ?$

  Ⓐ $90 - 2 - 8$
  Ⓑ $82 - (10 - 8)$
  Ⓒ $82 + (10 - 8)$
  Ⓓ $(80 + 2) - 8$

3. **Find the sum of the mixed numbers.**

$$4\frac{7}{8} + 7\frac{5}{8}$$

  Ⓐ $11\frac{5}{8}$

  Ⓑ $12\frac{1}{2}$

  Ⓒ $12\frac{7}{8}$

  Ⓓ $11\frac{1}{2}$

4. **Which is a correct use of the distributive property?**

  Ⓐ $3(4 \cdot 2) = 3(4) \cdot 3(2)$
  Ⓑ $3(4 \cdot 2) = 3(4) + 3(2)$
  Ⓒ $3(4 + 2) = 3(4) + 3(2)$
  Ⓓ All of the above

5. **Name the property illustrated in the equation:**

    (5 + 8) + 2 = 5 + (8 + 2)

    Ⓐ Distributive property
    Ⓑ Associative property of addition
    Ⓒ Commutative property of addition
    Ⓓ Identity property of addition

6. **Which is a valid use of properties to make the expression easier to calculate?**

    7 – 2.45

    Ⓐ (6 – 2) + (1 – 0.45)
    Ⓑ (7 – 2) + 0.45
    Ⓒ 5 + 0.45
    Ⓓ 7 – (0.45 – 2)

7. **Write the improper fraction as a mixed number.**

    $$\frac{31}{3}$$

    Ⓐ $3\frac{1}{3}$

    Ⓑ $30\frac{1}{3}$

    Ⓒ $31\frac{1}{3}$

    Ⓓ $10\frac{1}{3}$

8. **Find the difference.** $16\frac{3}{4} - 9\frac{7}{8}$

    Ⓐ $7\frac{1}{8}$

    Ⓑ $6\frac{7}{8}$

    Ⓒ $7\frac{7}{8}$

    Ⓓ $6\frac{1}{8}$

9. **What property is illustrated in the equation?:**

$(5 + 3) + 0 = 5 + 3$

Ⓐ Identity property of addition
Ⓑ Associative property of addition
Ⓒ Commutative property of addition
Ⓓ Distributive property

10. **Which is a valid use of properties to make the expression easier to calculate?**

$$\frac{20}{7} - 1\frac{3}{7} =$$

Ⓐ $\frac{20}{7} - 1 + \frac{3}{7}$

Ⓑ $\frac{13}{7} - 1 + \frac{3}{7}$

Ⓒ $(\frac{13}{7} - \frac{3}{7}) + (1 - 1)$

Ⓓ $\frac{20}{7} - \frac{1}{7} - \frac{3}{7}$

Rational Numbers, Multiplication and Division (7.NS.A.2.A)

1. **Fill in the blank to make a true equation.**

   (-9)(___) = 36

   Ⓐ - 4
   Ⓑ 4
   Ⓒ 6
   Ⓓ - 6

2. **Evaluate the following expression.**

   - 2(a + 3b) =

   Ⓐ - 2a + 3b
   Ⓑ - 2a - 3b
   Ⓒ - 2a + 6b
   Ⓓ - 2a - 6b

3. **Simplify the following complex fraction.**

   $$\dfrac{\dfrac{1}{2}}{\dfrac{2}{3}}$$

   Ⓐ $\dfrac{4}{3}$

   Ⓑ $\dfrac{1}{2}$

   Ⓒ $\dfrac{3}{4}$

   Ⓓ $\dfrac{4}{3}$

4. **Which of the following represents the product of 0.53 * 11.6?**

Ⓐ 0.6148
Ⓑ 6.148
Ⓒ 61.48
Ⓓ 614.8

5. **Which of the following is NOT a rational number?**

Ⓐ $\dfrac{8}{15}$

Ⓑ 20.6
Ⓒ $\sqrt{3}$

Ⓓ $-\dfrac{9}{11}$

6. **Which of the following numbers is divisible by 3?**

Ⓐ 459,732
Ⓑ 129,682
Ⓒ 1,999,000
Ⓓ 5,684,722

7. **Which of the following fractions is equivalent to 0.625?**

Ⓐ $\dfrac{7}{9}$

Ⓑ $\dfrac{5}{8}$

Ⓒ $\dfrac{2}{3}$

Ⓓ $\dfrac{8}{9}$

8. Mary wanted to estimate how many sheets of paper she had left. She knew that each sheet was about 0.1 mm thick. Her pack measured 2 cm thick.

   Which is the best estimate of the number of sheets of paper in Mary's pack?

   Ⓐ 50
   Ⓑ 100
   Ⓒ 150
   Ⓓ 200

9. Which of the following is the multiplicative inverse of (- 7/9)?

   Ⓐ 7/9
   Ⓑ - 9/7
   Ⓒ 9/7
   Ⓓ 1

10. Determine how many steps are necessary in a stairway between two floors that are 10.5 feet apart if each step is 9 inches high.

    Ⓐ 11 steps
    Ⓑ 12 steps
    Ⓒ 13 steps
    Ⓓ 14 steps

Rational Numbers As Quotients of Integers (7.NS.A.2.B)

1. **Which of the following division problems CANNOT be completed?**

    Ⓐ $10 \div 0$
    Ⓑ $155 \div (-3)$
    Ⓒ $\dfrac{2}{3} \div \dfrac{1}{4}$
    Ⓓ $0 \div 5$

2. **Which of the following is NOT equivalent to the given value?:**

    $-\dfrac{2}{3}$

    Ⓐ $\dfrac{-2}{(3)}$

    ● $\dfrac{-2}{(-3)}$

    Ⓒ $\dfrac{2}{(-3)}$

    Ⓓ All are equivalent values

3. **Jared hiked a trail that is 12 miles long. He hiked the trail in legs that were 1.5 miles each. How many legs did he hike?**

    Ⓐ 12
    Ⓑ 10
    Ⓒ 8
    Ⓓ 6

4. **Greg is able to run a mile in 8 minutes. He ran at that pace for t minutes. What does the following expression represent?**

    $\dfrac{t}{8}$

    Ⓐ The number of miles that Greg ran.
    Ⓑ The number of hours that Greg ran.
    Ⓒ The average pace at which Greg ran.
    Ⓓ The amount of time it took Greg to run 8 miles.

5. What value could x NOT be in the following expression?:

$$\frac{5+x}{x}$$

   Ⓐ 5
   Ⓑ 1
   Ⓒ –5
   Ⓓ 0

6. Rose is filling her swimming pool with water. She needs to pump 2000 gallons into the pool, and the water flows at a rate of r gallons per hour. Which of the following expresses the amount of time it will take to fill the pool?

   Ⓐ $\dfrac{2000}{r}$

   Ⓑ 2000*r*

   Ⓒ 2000 + *r*

   Ⓓ $\dfrac{r}{2000}$

7. If t = –2, and v = –4, which of the following is equal to $\dfrac{t}{v}$?

   Ⓐ $\dfrac{t}{4}$

   Ⓑ $\dfrac{4}{2v}$

   Ⓒ $\dfrac{-t}{4}$

   Ⓓ $\dfrac{-6}{-3v}$

8. If , $\dfrac{3a}{b} = 12$, what is the value of $-\dfrac{a}{b}$?

Ⓐ - 4

Ⓑ $\dfrac{3}{4}$

Ⓒ $-\dfrac{3}{4}$

Ⓓ $\dfrac{1}{4}$

9. **Which terms best complete the statement?**

    A number is rational if it can be expressed as a _____ of _____.

    Ⓐ sum; integers
    Ⓑ quotient; integers
    Ⓒ difference; prime numbers
    Ⓓ product; variables

10. **Kathy is laying stepping stones in her garden. The stones are 8 inches long, and she wants to create a path that is 10 feet long. How many stones will she need?**

    Ⓐ 10 stones
    Ⓑ 80 stones
    Ⓒ 15 stones
    Ⓓ 1.25 stones

Strategies for Multiplying and Dividing Rational Numbers (7.NS.A.2.C)

1. Find the product: $3 \times \dfrac{1}{3}$

   Ⓐ 1
   Ⓑ 3
   Ⓒ 9
   Ⓓ 6

2. Which property is illustrated in the following statement?

   $(5)(4)(7) = (7)(5)(4)$

   Ⓐ Triple multiplication property
   Ⓑ Distributive property
   Ⓒ Commutative property of multiplication
   Ⓓ Associative property of multiplication

3. Find the quotient: $4 \div \dfrac{1}{2}$

   Ⓐ 4
   Ⓑ 8
   Ⓒ 2
   Ⓓ 16

4. Which of the following is a helpful and valid way to rewrite the expression for evaluation :

   $3(123)$

   Ⓐ $3(100) + 3(20) + 3(3)$
   Ⓑ $3(100) + 23$
   Ⓒ $100 + 3(23)$
   Ⓓ $300 + 3(20) + 3$

5. **Which of the following is equivalent to the expression?:**

$6 (2 \dfrac{2}{3})$

Ⓐ $12 + \dfrac{2}{3}$

Ⓑ $12 \dfrac{2}{3}$

Ⓒ $\dfrac{12 + 2}{3}$

Ⓓ $6 (2) + 6 (\dfrac{2}{3})$

6. **Which property is illustrated in the following statement?**

$(2 \cdot 3) \cdot 6 = 2 \cdot (3 \cdot 6)$

Ⓐ Associate property of multiplication
Ⓑ Commutative property of multiplication
Ⓒ Identity property of multiplication
Ⓓ Reflexive property

7. **Find the product:** $\dfrac{3}{2} * \dfrac{7}{6}$

Ⓐ $\dfrac{10}{12}$

Ⓑ $\dfrac{21}{6}$

Ⓒ $\dfrac{7}{4}$

Ⓓ $\dfrac{10}{8}$

8. **Find the quotient:** $\dfrac{6}{5}$ + $\dfrac{3}{25}$

    (A) $\dfrac{2}{5}$

    (B)   2

    (C) $\dfrac{5}{2}$

    (D)   10

9. **Find the product:** $\left(2\,\dfrac{1}{3} \right)\left(3\,\dfrac{1}{2} \right)$

    (A) $\dfrac{49}{6}$

    (B) $6\,\dfrac{1}{6}$

    (C) $\dfrac{6}{5}$

    (D) $5\,\dfrac{2}{5}$

10. **Find the quotient:** $\left(5\,\dfrac{1}{4} \right) \div \left(1\,\dfrac{1}{2} \right)$

    (A) $\dfrac{5}{8}$

    (B) $\dfrac{7}{2}$

    (C)   45

    (D)   7

Converting Between Rational Numbers and Decimals (7.NS.A.2.D)

1. **Convert to a decimal:** $\dfrac{7}{8}$

    Ⓐ 0.78
    Ⓑ 0.81
    Ⓒ 0.875
    Ⓓ 0.925

2. **Convert to a decimal:** $\dfrac{5}{6}$

    Ⓐ 0.8333333…
    Ⓑ 0.56
    Ⓒ 0.94
    Ⓓ 0.8

3. **How can you tell that the following number is a rational number?**

    0.251

    Ⓐ It is a rational number because the decimal terminates.
    Ⓑ It is a rational number because there is a value of 0 in the ones place.
    Ⓒ It is a rational number because the sum of the digits is less than 10.
    Ⓓ It is a rational number because it is not a repeating decimal.

4. **A group of 11 friends ordered 4 pizzas to share. They divided the pizzas up evenly and all ate the same amount. Express in decimal form the portion of a pizza that each friend ate.**

    Ⓐ 0.36363636…
    Ⓑ 0.411
    Ⓒ 0.14141414…
    Ⓓ 0.48

5. **How can you tell that the following number is a rational number?**

    0.133333…

    Ⓐ It is rational because the decimal does not terminate.
    Ⓑ It is rational because the decimal repeats over and over.
    Ⓒ It is rational because the number is a factor of 1.
    Ⓓ It is NOT rational because the decimal does not terminate.

6. **Convert to a decimal:** $2\dfrac{2}{9}$

   Ⓐ 2.92299229…

   Ⓑ 2.2222…

   Ⓒ 2.35

   Ⓓ 2.4835

7. **How can you tell that the following number is not a rational number?:**

2.4876352586582142597868…

   Ⓐ It is not rational because the same digit never occurs twice in a row in the decimal.

   Ⓑ It is not rational because the decimal does not terminate or repeat.

   Ⓒ It is not rational because it is greater than 1.

   Ⓓ It is not rational because it is not a factor of 5.

8. **Convert to a decimal:** $\dfrac{11}{25}$

   Ⓐ 0.4444444…

   Ⓑ 0.472

   Ⓒ 0.369

   Ⓓ 0.44

9. **The track at Haley's school is a third of a mile in length. She ran 14 laps on it after school one day. Express the number of miles she ran in decimal form.**

   Ⓐ 4.56

   Ⓑ 3.85

   Ⓒ 4.6666…

   Ⓓ 4.725

10. **Convert to a decimal:** $\dfrac{34}{99}$

   Ⓐ 0.943

   Ⓑ 0.343434…

   Ⓒ 0.394

   Ⓓ 0.439439…

Solving Real World Problems (7.NS.A.3)

1. Andrew has $9.39 but needs $15.00 to make a purchase. How much more does he need?

    Ⓐ $6.39
    Ⓑ $5.61
    Ⓒ $5.39
    Ⓓ $6.61

2. Ben has to unload a truck filled with 25 bags of grain for his horses. Each bag weighs 50.75 pounds.

    How many pounds does he have to lift?

    Ⓐ 12,687.50 pounds
    Ⓑ 1,268.75 pounds
    Ⓒ 126.875 pounds
    Ⓓ 1250 pounds

3. A Chinese restaurant purchased 1528.80 pounds of rice. If they received 50 identical bags, how much rice was in each bag?

    Ⓐ 30.576 pounds
    Ⓑ 305.76 pounds
    Ⓒ 3.0576 pounds
    Ⓓ None of the above.

4. Leila stopped at the coffee shop on her way to work. She ordered 2 bagels, 3 yogurts, and 1 orange juice. Bagels were $0.69 each, yogurts were $1.49 each, and orange juice was $1.75. What was Leila's total bill?

    Ⓐ $7.60
    Ⓑ $3.93
    Ⓒ $5.16
    Ⓓ $5.42

5. Mickey bought pizza and sodas for himself and four of his friends. The pizza was $17.49, and 5 sodas were $1.19 each.

    If the pizza is sliced into 10 equal slices and each boy eats 2 slices and drinks one soda, what is the cost to each boy?

    Ⓐ $2.94
    Ⓑ $4.13
    Ⓒ $3.50
    Ⓓ $4.69

6. **Alan has to keep within a $15.00 budget. Tax is 6.5%. What is Alan's total if he buys 1 notebook, 1 pack of paper, 1 set of dividers, 2 pens and 5 pencils?**

3-Ring Notebook	$5.69
Notebook Paper	$1.39
Dividers	$1.45
Pens	$1.19
Pencils	$0.50

- Ⓐ $10.88
- Ⓑ $14.28
- Ⓒ $13.41
- Ⓓ $15.00

7. **Sammy is mowing the lawn. The lawn is 30 ft by 20 ft. Sammy cut a strip 5 ft by 10 ft and ran out of gas. How much more does he need to mow?**

- Ⓐ 250 sq ft
- Ⓑ 550 sq ft
- Ⓒ 50 sq ft
- Ⓓ 600 sq ft

8. **Dustin has leased 5 acres of land to raise produce for the farmers' market. He has already planted 5/8 of the land.**

How many more acres does he need to plant?

- Ⓐ 3 1/8 acres
- Ⓑ 4 acres
- Ⓒ 1 7/8 acres
- Ⓓ 3/8 acres

9. **Taylor bought a bag of marbles weighing 5.25 lb. Before he got to the car, the bag broke, spilling many marbles. To find out if he had recovered all of his marbles, he weighed the bag at home.**

He found that his bag of marbles now weighed 4.98 lb. What was the weight of the lost marbles?

- Ⓐ 0.27 lb
- Ⓑ 2.7 lb
- Ⓒ 0.173 lb
- Ⓓ 1.73 lb

10. Which of the following is a correct statement about the multiplication of two integers?

(A) If the signs are both negative, multiply the numbers and give the answer a negative sign.

(B) If the signs are both negative, multiply the numbers and give the answer a positive sign.

(C) If the signs are unlike, multiply the numbers and give the answer a positive sign.

(D) If the signs are unlike, multiply the numbers and give the answer the sign of the larger absolute value.

End of The Number System

The Number System

Answer Key
&
Detailed Explanations

Rational Numbers, Addition & Subtraction (7.NS.A.1.A)

Question No.	Answer	Detailed Explanation
1	C	Remember, a fraction can simply represent a ratio of shaded or selected objects to total number of objects. In this case, 1/2 means selecting 1 for every 2 objects in the picture. Since there are four shaded objects out of 8 total objects, that represents 4/8, which in simplest form is 1/2.
2	B	Remember: adding and subtracting rational numbers works just like integers. If you need to carry or borrow, the rules remain the same. 25 + 2.005 - 7.253 - 2.977 27.005 - 7.253 - 2.977 19.752 - 2.977 = 16.775
3	A	-3 4/5 + 9 7/10 - 2 11/20 = -3 16/20 + 9 14/20 - 2 11/20 -5 27/20 + 9 14/20 = -5 27/20 + 8 34/20 3 7/20 is the correct answer.
4	C	Don't forget that when subtracting, you must change the sign of the subtrahend and then proceed as in addition. The problem 268.75 - (- 46.99) then becomes 268.75 + (+ 46.99). Now we have 268.75 + 46.99. The correct answer is 315.74.
5	A	When a negative sign precedes parentheses, it must be distributed to each term inside parentheses. p - (p - q) becomes p - p + q. This becomes 0 + q, or q. p - (p - q) is the correct answer.
6	C	They spent 6 hours and 5 min total on their trip. Of that time, 4 hours and 50 minutes were spent driving. 4 5/6 hrs driving out of 6 1/12 hours total (29/6) out of (73/12) (29/6) (12/73) = 58/73 hours spent driving
7	B	If a = 5/6, b = -2/3 and c = - 1 1/3, find a - b - c. 5/6 - (- 2/3) - (- 1 1/3) = 5/6 + 4/6 + 1 2/6 = 1 11/6 =2 5/6 2 5/6 is the correct answer.
8	D	P = 2(length) + 2(width) P = 2(36) + 2(28) P = 72 + 56 = 128 ft 128 ft - 5 ft = 123 ft fencing needed 150 - 123 = 27 ft extra fencing
9	B	Starting with a full candy bar, we take away 1/2 leaving 1/2. Then we take away 1/2 of the remaining half leaving 1/4 of the original candy bar. Then we take away half of the 1/4 leaving 1/8 of the original bar. 1/8 is the correct answer.
10	C	3 1/2 - 1/3 = 3 3/6 - 2/6 = 3 1/6 cups 3 1/6 cups is the correct answer.

Question No.	Answer	Detailed Explanation
Add and Subtract Rational Numbers (7.NS.A.1.B)		
1	A	The fact that p + q is less than p tells us that q is negative. Because p = q is exactly 1/3 less than p, we know q is –1/3.
2	C	Anytime that you add a number and its opposite, the sum is always 0.
3	B	The difference between 12/5 and 4/5 will be the value of q. The difference is 8/5.
4	B	b is the length of the distance between 5/8 and 13/8. That difference is 8/8, or 1.
5	D	Because t and v have a sum of 0, we know that they are opposite values. If t is 5/2, then v is –5/2.
6	C	The distance between the two points is 6/5. We also see that the difference in their given values is 2n. Because adding 2n takes us to the left of the other point, we know that 2n is a negative value of 6/5. n must be half that value, or –3/5.
7	B	The sum of a number and its opposite is always 0.
8	C	It does not matter whether t is positive or negative, or whether t is greater than or less than r. If you add a negative number to t, you will always move left on the number line. So, t + r will be to the left of t because r is a negative number being added to t.
9	D	If point A is located at 3 on the number line and point B is located at (-2/3) then the distance between the points is three and two-thirds. |3 - (-2/3)| = 3 2/3.
10	D	There are two possibilities of points that are 9/4 away from 7/4. One of the points is 9/4 more than 7/4, which would be 16/4 (not an answer option). The other point is 9/4 less than 7/4, which is –2/4, or –1/2.
Additive inverse and distance between two points on a number line (7.NS.A.1.C)		
1	A	Both terms inside of the parentheses are being subtracted, and subtraction of a number is the same as addition of its opposite.
2	C	The distance between two points can be found by taking the absolute value of the difference of the two points.

Question No.	Answer	Detailed Explanation
3	B	Because Kyle and Mark traveled in opposite directions, let one of their distances be a negative value. The distance between the two of them is the absolute value of the difference of their positions, which is 8 when evaluated.
4	D	The form given is the absolute value of the difference of two numbers. Those two numbers could represent the coordinates of two points if the expression is the distance between them. G is located at 5, so H is likely located at −2.
5	B	Subtracting 3y is the same as adding the opposite of 3y: + (−3y)
6	A	Because they live in opposite directions from the park, let one of their distances from the park be a negative value. The distance between their homes, then, is the absolute value of the difference of the two distances.
7	D	The statement says that subtracting a number is the same as adding the opposite of that number. This is true for all real numbers.
8	C	The x in the expression is being added, not subtracted, so we do not want to replace it with addition of its opposite. The 3 inside parentheses is being subtracted, so we can replace it with addition of its opposite, −3.
9	D	The distance between two points can be represented in the form of the absolute value of the difference of the coordinates of the two points.
10	A	The distance between two points on a number line can always be represented in the form of the absolute value of the difference of the coordinates of the two points.

Strategies for adding and subtracting rational numbers (7.NS.A.1.D)

1	A	The commutative property states that parts of an addition statement can be written in any order and the same sum will result.
2	C	The 92 can be written as 82 + 10, so that the 10 and 8 can first be subtracted, and then the result can be added to 82.
3	B	The whole number parts can be added together, and the fraction parts can be added together, resulting in 11 and 12/8. The 12/8 part is a full 8/8, or 1, and an additional 4/8, so that we have 12 and 4/8, which simplifies to 12 and 1/2.

Question No.	Answer	Detailed Explanation
4	C	The distributive property only applies when there is addition or subtraction inside the parentheses and a number outside the parentheses being multiplied to what is inside the parentheses.
5	B	The associative property of addition states that different parts of an addition statement can be grouped together in any manner, and the same sum will still result.
6	A	The 7 can be written as 6 + 1, and the 2.45 can be written as (2 + 0.45). Subtract the 2 from the 6 and the 0.45 from the 1.
7	D	31/3 is the same as 30/3 + 1/3. 30/3 = 10, so we have 10 and 1/3.
8	B	Let the 16 be 15 + 1, and then subtract the 7/8 from the 1, leaving only 1/8. You now have 15 + 1/8 + 6/8 – 9. Subtract the 9 from the 15, and add the fractions to arrive at 6 and 7/8.
9	A	The identity property of addition states that if you add 0 to a value, the sum is the same as the value itself.
10	C	Let 20/7 be 13/7 + 7/7. Subtract the 3/7 from the 13/7, and let the 7/7 be written as 1 and subtracted with the existing 1.

Rational Numbers, Multiplication and Division (7.NS.A.2.A)

1	A	Remember that when we multiply like signs, either (- * -) or (+ * +), the result is positive, and when we multiply unlike signs, the result is negative. Since we are multiplying some number by - 9 and get +36, we know that the sign of the unknown quantity is also negative. Now, 4 * 9 = 36; so - 4 * - 9 = 36. - 4 is the correct answer.
2	D	This problem requires the use of the distributive property. We must distribute the quantity outside parentheses to each term inside parentheses. - 2(a + 3b) = - 2a - 6b. - 2a - 6b is the correct answer.
3	C	In order to solve 1/2 ÷ 2/3. Keep the first fraction, change the division symbol into a multiplication symbol, and flip the second fraction before solving the problem. (1) 1/2 ÷ 2/3 = (2) 1/2 × 3/2 = (3) (1 × 3)/(2 × 2) = (4) 3/4 3/4 is the correct answer.

Name: _____   Date: _____

Question No.	Answer	Detailed Explanation
4	B	Multiply the numbers, and then count the total number of decimal places in the multiplier and the multiplicand. Start at the right, and count back the total number of places in the multiplier and multiplicand. Place the decimal there. 0.53 has 2 decimal places, and 11.6 has 1. 2 + 1 = 3. Therefore, we count back 3 places. The product gave the numbers 6148. We count back 3 places and the decimal falls between 6 and 1. 6.148 is the correct answer.
5	C	Rational numbers are numbers that can be written as p/q where q is not 0. To remember this definition, associate rational with ratio. $\sqrt{3}$ is the correct answer.
6	A	If the sum of the digits is divisible by 3, then the number is divisible by 3. The sum of the digits of 459,732 is 30 which is divisible by 3. Therefore, 459,732 is divisible by 3. 459,732 is the correct answer.
7	B	0.625 = 625/1000 Now divide top and bottom by 125 to reduce to lowest terms. 625/1000 = 5/8. 5/8 is the correct answer.
8	D	We must first make the units of measure the same. The prefix milli means 1/1000 and the prefix centi means 1/100. 10(1/1000) = 10/1000 = 1/100. Therefore, 10 millimeters (mm) = 1 centimeter (cm). Then 2 cm = 2(10) mm = 20 mm. 20 mm/ 0.1 mm = 200 Mary has approximately 200 sheets of paper in her pack. 200 is the correct answer.
9	B	Multiplicative inverse is another name for reciprocal. When you multiply a number by its multiplicative inverse, you get 1. (- 7/9) x (- 9/7) = + 63/63 = + 1. The multiplicative inverse of (- 7/9) is (- 9/7). (- 9/7) is the correct answer.
10	D	Change feet to inches so you are working with the same units. 10.5 ft = 12(10.5) = 126 inches. Now, we must determine the number of 9 inch steps by dividing the total distance between floors, 126 inches, by the height of each step, 9 inches. 126/9 = 14 steps 14 is the correct answer.

Rational numbers as quotients of integers (7.NS.A.2.B)

1	A	0 cannot be the divisor in a division problem. 0 cannot be the divisor in a division problem because division by 0 is undefined.
2	B	Dividing a negative number by a negative number results in a positive number. If only one of the dividend or divisor is negative, then the quotient is negative.

Question No.	Answer	Detailed Explanation
3	C	Divide the length of the trail by the length of one leg, and the number of legs results: 8.
4	A	Dividing the total time Greg ran by the time per mile will result in the total number of miles Greg ran.
5	D	The divisor in a division problem cannot be 0.
6	A	Divide the total amount of water that must be pumped into the pool by the number of gallons per hour, and the result will be the number of hours required to fill the pool.
7	C	In the fraction (-t)/4, the numerator is the opposite of t, and the denominator is the opposite of v. The fact that both numerator and denominator are opposites will result in a cancelling of the two negatives, and the value will be the same as the original.
8	A	We can see that a/b must have a value of 4 in order for 3a/b to have a value of 12. The opposite of a/b, then, must be −4.
9	B	The definition of a rational number is that it can be expressed as a ratio (or quotient) of integers.
10	C	Divide the total length in inches (120) by the number of inches per stone, and the quotient will be the number of stones needed, 15.

Strategies for multiplying and dividing rational numbers (7.NS.A.2.C)

1	A	Multiply the 3 by the numerator, 3, and then divide by the denominator of 1. The result is 1.
2	C	The commutative property states that multiplication can be done in any order and the same product will result.
3	B	Dividing by a fraction is the same thing as multiplying by the reciprocal of that fraction. Multiply, then, by 2 rather than dividing by 1/2: 4 x 2 = 8.
4	A	The 3 must be multiplied by all three place values. The same product will result if writing it out in individual terms.
5	D	The mixed number is a sum of the whole number part and the fraction part. Distributing the multiplication to both terms results in the same value.
6	A	The associative property of multiplication states that multiplication can be grouped in any way and the same product will result.

Question No.	Answer	Detailed Explanation
7	C	Multiply the numerators together and the denominators together. This results in a value of 21/12, which simplifies to 7/4.
8	D	Instead of dividing by 3/25, multiply by the reciprocal, 25/3. The result is 150/15, which simplifies to 10.
9	A	Convert both mixed numbers to improper fractions, and multiply them: (7/3)(7/2) = 49/6.
10	B	Convert both fractions to *rational* numbers (21/4)÷(3/2) and multiply by the reciprocal (21/4) • (2/3) equals 42/12. 42/12 reduces to 7/2.

Converting between rational numbers and decimals (7.NS.A.2.D)

1	C	Divide 7 by 8 using long division, and the result is 0.875.
2	A	Divide 5 by 6 using long division. The decimal repeats endlessly: 0.83333...
3	A	If the decimal terminates, then the number can always be written as the quotient of two integers and is rational.
4	A	Divide 4 by 11 using long division. The decimal repeats endlessly: 0.363636...
5	B	If the decimal repeats endlessly, the number can be expressed as a quotient of integers and is rational.
6	B	The whole number part is the value of the ones digit. Divide the fraction to learn the decimal part of the number. The result is a repeating decimal: 2.2222...
7	B	Any decimal number that continues endlessly without repeating is not a rational number.
8	D	Divide the numerator by the denominator using long division. The result is 0.44.
9	C	Divide the number of laps by 3 to find how many miles she ran. The result is a repeating decimal: 4.6666...
10	B	Divide the numerator by the denominator using long division. The result is a repeating decimal: 0.343434...

Question No.	Answer	Detailed Explanation
		Solving Real World Problems (7.NS.A.3)
1	B	$15.00 - $9.39 = $15.00 - $ 9.39 $ 5.61 is the correct answer.
2	B	50.75 x 25 = 1268.75 pounds 1268.75 pounds is the correct answer.
3	A	1528.80 pounds ÷ 50 bags = 30.576 pounds in each bag. 30.576 is the correct answer.
4	A	2 bagels @ $0.69 ea = 2 x $0.69 = $1.38 3 yogurts @ $1.49 ea = 3 x $1.49 = $4.47 1 orange juice @ $1.75 = $1.75 Total bill = $7.60 $7.60 is the correct answer.
5	D	The pizza was divided equally so each boy's cost for pizza is $17.49 ÷ 5 = $3.50. Soda was $1.19 each. Add to find the total cost for each boy. $3.50 + $1.19 = $4.69. $4.69 is the correct answer.
6	B	1 notebook = $5.69 1 pack paper = $1.39 1 set dividers = $1.45 2 pens = 2 x $1.19 = $2.38 5 pencils = 5 x $0.50 = $2.50 Subtotal = $13.41 Tax = .065 x $13.41 = $0.87 Total = $14.28 The answer is $14.28.
7	B	The lawn is 30 ft x 20 ft = 600 sq ft. He mowed 5 ft x 10 ft = 50 sq ft 600 sq ft - 50 sq ft = 550 sq ft 550 sq ft is the correct answer.
8	C	5/8 x 5 = 25/8 = 3 1/8 acres already planted. 5 - 3 1/8 = 4 8/8 - 3 1/8 = 1 7/8 acres remaining to plant. 1 7/8 acres is the correct answer.
9	A	5.25 - 4.98 = 0.27 lb 0.27 lb is the correct answer.
10	B	When multiplying or dividing two integers, like signs will always give a positive, and unlike signs will always give a negative sign. If the signs are both negative, multiply the numbers and give the answer a positive sign, is the correct answer.

Expressions and Equations

1. Ruby is two years younger than her brother. If Ruby's brother's age is A, which of the following expressions correctly represents Ruby's age?

    Ⓐ A - 2
    Ⓑ A + 2
    Ⓒ 2A
    Ⓓ 2 - A

2. Find the difference: 8n - (3n - 6) =

    Ⓐ -n
    Ⓑ 5n - 6
    Ⓒ 5n + 6
    Ⓓ 8n - 6

3. Find the sum:

    6t + (3t - 5) =

    Ⓐ 9t - 5
    Ⓑ 9t + 5
    Ⓒ 3t - 5
    Ⓓ 6t - 5

4. Combine like terms and factor the following expression.

    7x - 14x + 21x - 2

    Ⓐ 14x - 2
    Ⓑ 2(7x - 1)
    Ⓒ 42x - 2
    Ⓓ 21(x - 1)

5. **Which of the following expressions is equivalent to:**

   3(x + 4) - 2

   Ⓐ 3x + 10
   Ⓑ 3x + 14
   Ⓒ 3x + 4
   Ⓓ 3x + 5

6. **Simplify the following expression:**

   $(\dfrac{1}{2}) x + (\dfrac{3}{2}) x$

   Ⓐ 2x

   Ⓑ $(\dfrac{5}{2}) x$

   Ⓒ - x

   Ⓓ $\dfrac{1}{2}$

7. **Simplify the following expression:**

   0.25x + 3 - 0.5x + 2

   Ⓐ -0.25x + 5
   Ⓑ 0.75x + 5
   Ⓒ -0.25x + 1
   Ⓓ 5.75x

8. **Which of the following statements correctly describes this expression?**

   2x + 4

   Ⓐ Four times a number plus two
   Ⓑ Two more than four times a number
   Ⓒ Four more than twice a number
   Ⓓ Twice a number less four

9. **Which of the following statements correctly describes the following expression?**

$$\frac{2x - 3}{2}$$

Ⓐ Half of three less than twice a number
Ⓑ Half of twice a number
Ⓒ Half of three less than a number
Ⓓ Three less than twice a number

10. **Which of the following expressions is not equivalent to:**

$$(\frac{1}{2}) (2x + 4) - 3$$

Ⓐ $(x + 4) - 3$

Ⓑ $(\frac{1}{2}) (2x + 4) - 3$

Ⓒ $x - 1$

Ⓓ $x + 1$

11. **Which property is demonstrated in the following expression?**

$$12(3x - 9) = 36x - 108$$

Ⓐ Associative property
Ⓑ Distributive property
Ⓒ Identity property of addition
Ⓓ Zero property of multiplication

12. **Rhonda is purchasing fencing to go around a rectangular lot which is 4x + 9 ft long and 3x - 5 ft wide. Which expression represents the amount of fencing she must buy?**

Ⓐ $7x + 4$
Ⓑ $7x - 4$
Ⓒ $14x + 28$
Ⓓ $14x + 8$

13. **Rebekah is preparing for a swim meet. She is trying to swim 1 mile in 7 minutes. If the pool is 5x + 3 ft long, which expression represents how many laps she needs to swim in 7 minutes?**

    Assume 1 length of the pool is 1 lap. (1mile = 5280 feet.)

    Ⓐ 7(5x + 3)
    Ⓑ 5x + 3
    Ⓒ 5280 / (5x + 3)
    Ⓓ 5280(5x + 3)

14. **Simplify.**

    5x + 10y + 0(z) =

    Ⓐ 0
    Ⓑ 5x + 10y
    Ⓒ 15xy
    Ⓓ 5x

15. **Which property is demonstrated below?**

    2 + (8 + 3) = (2 + 8) + 3

    Ⓐ Additive Identity Property
    Ⓑ Multiplicative Identity Property
    Ⓒ Distributive Property
    Ⓓ Associative Property of Addition

Interpreting the Meanings of Expressions (7.EE.A.2)

1. Which of the following expressions represents "5% of a number"?

    Ⓐ 5n
    Ⓑ 0.5n
    Ⓒ 0.05n
    Ⓓ 500n

2. Jill is shopping at a department store that is having a sale this week. The store has advertised 15% off certain off-season merchandise. Jill calculates the sales price by multiplying the regular price by 15% and then subtracting that amount from the regular price: SP = RP - 0.15(RP), where S = Sales Price and R = Regular Price. Find a simpler way for Jill to calculate the sales price as she shops.

    Ⓐ SP = 0.15 RP
    Ⓑ SP = 1.15RP
    Ⓒ SP = 0.85RP
    Ⓓ SP = 1.85RP

3. Rewrite the following expression for the perimeter of a rectangle.

    P = l + w + l + w, where P = perimeter, l = length, and w = width.

    Ⓐ P = l + 2w
    Ⓑ P = l + w
    Ⓒ P = 2(l) + w
    Ⓓ P = 2(l + w)

4. If an item costs a store x dollars to buy and can be sold at y dollars, what percentage of the sale price is profit?

    percentage of the sale price = [profit ÷ sale price]100%

    Which of the following rewrites this expression and includes the given information?

    Ⓐ (y - x) / 100y
    Ⓑ 100(y - x) / y
    Ⓒ (x - y) / 100y
    Ⓓ 100(x - y) / y

5. **To find the average of five consecutive integers beginning with x, we add the integers and divide by 5.**

   Expressed in mathematical symbols, we have (x + x + 1 + x + 2 + x + 3 + x + 4) / 5 = Average (a)

   Rewrite this expression another way.

   Ⓐ x + 4
   Ⓑ x + 3
   Ⓒ x + 2
   Ⓓ None of the above.

6. **The bank will charge a 10% overdraft fee for any money withdrawn over the amount available in an account. If Jared has d dollars in the bank and he withdraws (d + 5) dollars, the bank charges 0.10{(d + 5) - d}. Find a simpler way of rewriting his overdraft fee.**

   Ⓐ 0.10(d + 5)
   Ⓑ $0.50
   Ⓒ $5.00
   Ⓓ $2.50

7. **Bob has been logging the days that the temperature rises over 100° in Orlando. He found that on 18 of the past 30 days the temperature rose above 100°.**

   Using this information, he made the following prediction of the number of days out of the next n days that the temperature would NOT rise above 100°:

   P = [(30-18) / 30]n days. Find another way to write this expression.

   Ⓐ (2/5) n days
   Ⓑ (3/5) n days
   Ⓒ (4/5) n days
   Ⓓ n days

8. **If Roni runs for 20 minutes, walks for 10 and then runs for 15 while covering a total distance of m miles, her rate would be represented by:**

   r = m /{(20 + 10 +15) / 60} mph. Which of the following is a simpler representation of this formula?

   Ⓐ r =(m/45) mph
   Ⓑ r = (4/3m) mph
   Ⓒ r = (45m/60) mph
   Ⓓ r = (4m/3) mph

9. **Hannah is taking piano lessons. Her mother has told her that each week she takes lessons, she has to practice 1/2 hour more than she did the week before. Hannah created the following table. Write a simple expression to find the time she will practice in week 5n?**

Piano Practice Time Sheet

Week	Time
1	1
2	1 1/2
n	1 + (1/2) (n-1)

Ⓐ (1 + 5n)/2 hr.
Ⓑ (5n)/2 hr.
Ⓒ (5n - 1)/2 hr.
Ⓓ 1 + (5n/2) hr.

10. **The diameter of the Sun at the equator is 1,400,000 km. d = 1,400,000 km**

Which of the following is another way to write this expression?

Ⓐ $d = 1.4 \times 10^6$ km
Ⓑ $d = 1.4 \times 10^{-6}$ km
Ⓒ $d = 14 \times 10^6$ km
Ⓓ $d = 14 \times 10^{-6}$ km

11. **Which is another way of writing the following mathematical expression?**

(6x - 15y - 2x + y) / -4(x - y)

Ⓐ (2x - 7y) /(-2x -2y)
Ⓑ (2x - 7y) / -2
Ⓒ 5/2
Ⓓ (2x - 7y) / - 2(x - y)

12. **Which of the following expressions represents "a 24% increase"?**

Ⓐ n + 0.24
Ⓑ n + 0.24n
Ⓒ n + 24
Ⓓ n + 24n

13. Ted is monitoring the scores of his two favorite football teams. He has a bet that Team A will finish with 19 points ahead of Team B.

He has made the following statement: Team A score -19 = Team B score. Which of the following is another way of making the same statement?

Ⓐ Team B Score - 19 points = Team A Score
Ⓑ Team B Score / Team A Score = 19 points
Ⓒ Team A Score + Team B Score = 19 points
Ⓓ Team A Score = Team B Score + 19 points

14. Linda bought 3 packages of red/white/blue ribbon for a rally. Each package held 3 rolls at 50 yd long and 3 rolls at 25 yd long, and each package cost $6.99.

Cost / yd = [3($6.99)] / 3[3(50) + 3(25)]

Which of the following is a simpler way of expressing cost/yd of the ribbon?

Ⓐ $0.031/yd
Ⓑ $0.31/yd
Ⓒ $3.10/yd
Ⓓ $2.10/yd

15. Jack went to the feed store to buy grain for his livestock. He bought fifty-five 50 lb bags.

Cost = $40/bag, but for every 10 bags, he got 1 free.

The following mathematical expression shows the cost/lb.

Cost/lb=$40(50)/[50(50) + 5(50)]

Which of the following is a simpler way of writing the above expression?

Ⓐ Cost/lb = $50(50) / [40(55)]
Ⓑ Cost/lb = $50(50) / [40(50)]
Ⓒ Cost/lb = $40(55) / [50(50)]
Ⓓ Cost/lb = $40(50) / [55(50)]

Modeling Using Equations or Inequalities (7.EE.B.3)

1. **A 30 gallon overhead tank was slowly filled with water through a tap. The amount of water (W, in gallons) that is filled over a period of t hours can be found using W = 3.75(t). If the tap is opened at 7 AM and closed at 3 PM, how much water would be in the tank? Assume that the tank is empty before opening the tap.**

    Ⓐ 18 gallons
    Ⓑ 20 gallons
    Ⓒ 24 gallons
    Ⓓ The tank is full

2. **The ratio (by volume) of milk to water in a certain solution is 3 to 8. If the total volume of the solution is 187 cubic feet, what is the volume of water in the solution?**

    Ⓐ 130 cubic feet
    Ⓑ 132 cubic feet
    Ⓒ 134 cubic feet
    Ⓓ 136 cubic feet

3. **A box has a length of 12 inches and width of 10 inches. If the volume of the box is 960 cubic inches, what is its height?**

    Ⓐ 6 inches
    Ⓑ 10 inches
    Ⓒ 12 inches
    Ⓓ 8 inches

4. **Jan is planting tomato plants in her garden. Last year she planted 24 plants and harvested 12 bushels of tomatoes during the season. This year she has decided to only plant 18 plants. If the number of plants is directly proportional to the number of bushels of tomatoes harvested, how many bushels of tomatoes should she expect this year?**

    Ⓐ 6
    Ⓑ 18
    Ⓒ 12
    Ⓓ 9

5. **Melanie's age added to Roni's age is 27. Roni's age subtracted from Melanie's age is 3. Find their ages.**

    Ⓐ 17, 10
    Ⓑ 16, 11
    Ⓒ 15, 12
    Ⓓ 14, 13

6. Tim wraps presents at a local gift shop. If it takes 2.5 meters of wrapping paper per present, how many can Tim wrap if he has 50 meters of wrapping paper?

   Ⓐ 18 meters
   Ⓑ 20 meters
   Ⓒ 15 meters
   Ⓓ 17 meters

7. Name the property demonstrated by the equation.

   11 + (8 + 6) = (y + 8) + 6

   Ⓐ 11, Commutative Property of Addition
   Ⓑ 11, Associative Property of Addition
   Ⓒ 11, Distributive Property
   Ⓓ 11, Associative Property of Multiplication

8. In the linear equation p = 2c + 1, c represents the number of couples attending a certain event, and p represents the number of people at that event.

   If there are 7 couples attending, how many people will be present?

   Ⓐ 7 people
   Ⓑ 14 people
   Ⓒ 15 people
   Ⓓ 16 people

9. The ratio (by volume) of salt to sugar in a certain mixture is 4 to 8. If the total volume of the solution is 300 cubic feet, what is the volume of sugar in the mixture?

   Ⓐ 200 cubic feet
   Ⓑ 199 cubic feet
   Ⓒ 201 cubic feet
   Ⓓ 136 cubic feet

10. Which of the following sequences follows the rule $(8 + t^2) - 2t$ where t is equal to the number's place in the sequence?

   Ⓐ 7, 8, 11, 16, 23, ...
   Ⓑ 9, 12, 17, 24, 33 ...
   Ⓒ 3, 4, 5, 6, 7, ...
   Ⓓ 7, 10, 15, 22, 31, ...

Solving Multi-Step Problems (7.EE.B.4.A)

1. Bob, the plumber, charges 1/4 the cost of materials as his labor fee. If his current job has a material cost of $130, how much will Bob charge his client (including his labor fee)?

   Ⓐ $162.50
   Ⓑ $32.50
   Ⓒ $130.25
   Ⓓ None of the above

2. A box has a length of 6x inches. The width equals one third the length, and the height equals half the length. If the volume equals 972 cubic inches, what does x equal?

   Ⓐ 5
   Ⓑ 2
   Ⓒ 3
   Ⓓ 4

3. Taylor is trimming the shrubbery along three sides of his backyard. Their backyard is rectangular in shape, and the shrubbery lines one length and two widths of that rectangle. The length is twice the width and the total perimeter of the backyard is 180 feet. Find the length of the shrubbery that Taylor has to trim.

   Ⓐ 180 ft
   Ⓑ 60 ft
   Ⓒ 90 ft
   Ⓓ 120 ft

4. Jim is 4 years older than his brother Bob. In two years, Jim will be twice Bob's age. How old are they now?

   Ⓐ Bob is 6 and Jim is 10.
   Ⓑ Bob is 4 and Jim is 8.
   Ⓒ Bob is 0 and Jim is 4.
   Ⓓ Bob is 2 and Jim is 6.

5. In a certain classroom, the ratio of boys to girls is 2 to 1. If there are 39 students in the classroom, how many are boys?

    Ⓐ 18
    Ⓑ 22
    Ⓒ 26
    Ⓓ 30

6. John put three gallons of gasoline into his truck. The gasoline level was at 10% before he added the gasoline. If the truck has a 12 gallon tank, how much more gasoline can fit in the tank?

    Ⓐ 7.8 gallons
    Ⓑ 6.7 gallons
    Ⓒ 8.9 gallons
    Ⓓ 10.8 gallons

7. Melanie has $35.00 in her savings account and is working at a neighbor's house cleaning for $15.00 per week. Sue has no money saved, but is mowing lawns at $20.00 each. If Sue mows 1 lawn per week, how long will it take her to catch up with Melanie?

    Ⓐ 5 weeks
    Ⓑ 6 weeks
    Ⓒ 7 weeks
    Ⓓ 8 weeks

8. Sissy is baking cookies for her class party. She plans to bake 128 cookies. Her recipe makes 6 dozen cookies. If her recipe calls for 3 1/2 c flour, how much flour will she need to make 128 cookies (round to the nearest half cup)?

    Ⓐ 5 1/2 c
    Ⓑ 4 c
    Ⓒ 4 1/2 c
    Ⓓ 6 c

9. Jenn went to the farmer's market with $40.00. She bought a 10-lb bag of potatoes for $6.00, a pie for $8.00, 4 qt fresh blueberries for $4.00 per qt, and 5 lb of apples at $1.49 per lb. What percent of the $40.00 did she still have when she left?

    Ⓐ 93.625%
    Ⓑ 6.375%
    Ⓒ 25%
    Ⓓ .0595%

10. Nelly spent 7/8 of his savings on furniture and the rest on a lawnmower. If the lawnmower cost him $250, how much did he spend on furniture?

- (A) $2000
- (B) $1750
- (C) $1500
- (D) $1000

Linear Inequality Word Problems (7.EE.B.4.B)

1. The annual salary for a certain position depends upon the years of experience of the applicant. The base salary is $50,000, and an additional $3,000 is added to that per year of experience, y, in the field. The company does not want to pay more than $70,000 for this position, though. Which of the following inequalities correctly expresses this scenario?

    Ⓐ $53,000y \leq 70,000$
    Ⓑ $3,000y \leq 50,000$
    Ⓒ $50,000 + 3,000y \leq 70,000$
    Ⓓ $3,000 + 50,000y \leq 70,000$

2. Huck has $225 in savings, and he is able to save an additional $45 per week from his work income. He wants to save enough money to have at least $500 in his savings. Express this situation as an inequality.

    Ⓐ $265w \geq 500$
    Ⓑ $225 + 45w \geq 500$
    Ⓒ $225 \leq 45w$
    Ⓓ $225 + 45w \leq 500$

3. Lucy is charging her phone. It has a 20% charge right now and increases by an additional 2% charge every 3 minutes. She doesn't want to take it off the charger until it is at least 75% charged. Express this situation as an inequality.

    Ⓐ $20 + \dfrac{2}{3}\, m \leq 75$

    Ⓑ $20m + \dfrac{2}{3}\, m \leq 75$

    Ⓒ $75 + \dfrac{2}{3}\, m \geq 20$

    Ⓓ $20 + \dfrac{2}{3}\, m \geq 75$

4. Matt's final grade, G, in class depends on his last test score, t, according to the expression G = 74 + 0.20t. If he wants to have a final grade of at least 90.0, what is the minimum score he can make on the test?

   Ⓐ 80
   Ⓑ 74
   Ⓒ 92
   Ⓓ 86

5. Students can figure out their grade, G, on the test based upon how many questions they miss, n. The formula for their grade is G = 100 – 4n. If Tim wants to make at least an 83, what is the most questions he can miss?

   Ⓐ 5 questions
   Ⓑ 3 questions
   Ⓒ 4 questions
   Ⓓ 6 questions

6. The necessary thickness, T, of a steel panel in mm depends on the unsupported length, L, of the panel in feet according to the inequality T ≥ 3 + 0.2L. If the thickest panel available has a thickness of 12 mm, what is the maximum length it can span?

   Ⓐ 75 ft
   Ⓑ 45 ft
   Ⓒ 60 ft
   Ⓓ 54 ft

7. The pay that a salesperson receives each week is represented by the inequality P ≥ 300 + 25s, where s represents the number of units the salesperson sells. What is the significance of the number 300 in this inequality in this context?

   Ⓐ The salesperson will never make more than $300 weekly.
   Ⓑ The salesperson has never sold more than 300 units in a week.
   Ⓒ The salesperson is guaranteed at least $300 even if he doesn't sell any units.
   Ⓓ $300 is the price of a single unit.

8. The net calories gained or lost by a dieter depends on the hours exercised in a week according to the equation C = 750 – 200h. Based on his goals, Neil determines that the following inequality is necessary for him: h > 3.75. What is the significance of the value of this inequality?

   Ⓐ Neil will lose at least 3.75 pounds per week if he meets this inequality.
   Ⓑ Neil will lose weight each week if he meets this inequality.
   Ⓒ Neil can eat only 375 calories in any given day.
   Ⓓ If Neil meets this inequality, he will be able to eat as many calories as he likes.

9. A school always allows twice as many girls as boys enroll in swimming classes each year. They also follow a certain inequality for the number of boys, B, and girls, G, in their admissions: B + G ≤ 300. What significance does this inequality have for the number of boys that can be admitted each year?

Ⓐ At least 200 boys must be admitted.
Ⓑ No more than 100 boys can be admitted.
Ⓒ The number of boys admitted must equal the number of girls.
Ⓓ No more than 150 boys can be admitted.

10. Isaac has to do at least 4 hours of chores each week. For each hour of television he watches, that minimum number of hours increases by 0.25. This week, the inequality for his hours of chores is h ≥ 5.5. What does this indicate about his television watching?

Ⓐ He watched 6 hours of television this week.
Ⓑ He watched 5.5 hours of television this week.
Ⓒ He watched 22 hours of television this week.
Ⓓ He watched no television this week.

End of Expressions and Equations

Expressions and Equations

Answer Key
&
Detailed Explanations

Applying Properties to Rational Expressions (7.EE.A.1)

Question No.	Answer	Detailed Explanation
1	A	Remember: "Younger than" is a key phrase that will indicate subtraction. If A is the age of Ruby's brother, and she is 2 years younger than her brother, the correct expression is A - 2.
2	C	8n - (3n-6) = 8n - 3n + 6 = 5n + 6 5n + 6 is the correct answer.
3	A	6t + (3t - 5) = Remove parentheses: 6t + (3t − 5) = 6t + 3t - 5. Now combine like terms: 6t + 3t − 5 = 9t − 5 9t - 5 is the correct answer.
4	B	7x - 14x + 21x - 2 We can factor first or combine like terms first. Here, I will combine like terms first. 7x - 14x + 21x = 14x. Now we have 14x − 2, which factors into 2(7x - 1). 2(7x - 1) is the correct answer.
5	A	Simplify the expression 3(x + 4) - 2 Step 1: 3(x + 4) - 2 Step 2: 3x + 12 - 2 Step 3: 3x + 10
6	A	Simplify the expression 1/2x + 3/2x Step 1: 1/2x + 3/2x Add the numerators of the fractions and keep the same denominators. Step 2: 4/2x Step 3: 2x
7	A	Simplify the expression 0.25x + 3 - 0.5x + 2 by combining like terms Step 1: 0.25x + 3 - 0.5x + 2 Step 2: 0.25x - 0.5x + 3 + 2 Step 3: -0.25x + 5
8	C	Remember: Addition means more than and 2x represents multiplication, or times. The statement that represents the expression 2x + 4 is four more than two times a number.
9	A	Remember: A half can be represented by the fraction 1/2, twice indicates multiplication, and less represents subtraction. The statement that represents the expression (2x - 3) 2 is half of three less than twice a number because x is multiplied by 2 and the product is subtracted by three.
10	D	To find out which number is not equivalent simplify the expression: Step 1: x + 2 - 3 Step 2: x - 1 All of the answer choices will simplify to the same value except x + 1, which is not equivalent to x − 1.
11	B	Distributive property is the correct answer. To remove parentheses, we must distribute the quantity outside parentheses to each term inside parentheses.

Question No.	Answer	Detailed Explanation
12	D	Perimeter = 2 • length + 2 • width 2(4x + 9) + 2(3x - 5) = 8x + 18 + 6x - 10 = 14x +8 ft 14x + 8 is the correct answer.
13	C	5280 ft ÷ (5x + 3) ft represents the number of laps in 1 mile. 5280 / (5x + 3) is the correct answer.
14	B	5x + 10y + 0(z) = 5x + 10y When 0, the multiplicative identity, is multiplied by any quantity, it produces 0; so 0(z) = 0. Then 5x + 10y + 0 = 5x + 10y 5x + 10y is the correct answer.
15	D	The associative property of addition states that the sum of 3 or more numbers is the same regardless how the numbers are grouped.

Interpreting the Meanings of Expressions (7.EE.A.2)

1	C	A percent is a ratio that compares a number to 100. Therefore, 5% is 5 out of 100. To convert a percent to a decimal, divide by 100: 5 ÷ 100 = 0.05. To find 5% of a number, multiply the number times the decimal form of 5%: 0.05n. 0.05n is the correct answer.
2	C	SP = RP - 0.15(RP) SP = RP(1 - 0.15) SP = RP(0.85) SP = 0.85RP is the correct answer.
3	D	P = l + w + l + w Combine like terms. P = 2(l) + 2(w) Factor out 2. P = 2(l + w) is the correct answer.
4	B	Percentage of the sale price = [profit ÷ sale price]100 = [(y - x) ÷ y]100 = 100(y - x)/y is the correct answer.
5	C	(x + x + 1+ x + 2 + x + 3 + x + 4) / 5 = Average (a) (5x + 10) / 5 = x + 2 x + 2 is the correct answer.
6	B	0.10 ($ (d + 5) - $ d) 0.10 (d + 5 - d) 0.10(5) $0.50 is the correct answer.
7	A	[(30-18) / 30] n = (12/30) n = (2/5) n days is the correct answer.
8	D	r = m /[(20 + 10 +15) / 60] r = m / (45/60) = m / (3/4) = m (4/3) = 4m/3 mph
9	A	During week 5n, Hannah must practice 1 + 1/2(5n -1). 1 + (5n - 1)/2 = 2/2 + (5n - 1)/2 = (2 + 5n - 1)/2 = (1 + 5n) / 2 is the correct answer.
10	A	Scientific notation is a simpler way to compute with very large or very small numbers. Using scientific notation, 1,400,000 km = 1.4×10^6 km 1.4×10^6 km is the correct answer.

Name: _____ Date: _____

Question No.	Answer	Detailed Explanation
11	D	(6x - 15y - 2x + y) / -4(x - y) (4x - 14y) / -4(x - y) 2(2x - 7y) / -4(x - y) (2x - 7y) / - 2(x - y) is the correct answer.
12	B	If n is any number, and that number is increased by 24%, finding that increase means multiplying that number by 24% and then adding the product to the original number. Therefore, the expression n + 0.24n represents a 24% increase.
13	D	Team A score -19 = Team B score Add 19 to both sides of the equation. Team A Score -19 + 19 = Team B Score + 19 Team A Score = Team B Score + 19 is the correct answer.
14	A	Cost / yd = [3($6.99)] / 3[3(50) + 3(25)] $20.97/675 yd $0.031/yd is the correct answer.
15	D	Cost/lb=$40(50)/[50(50) + 5(50)] Cost/lb =[$40(50)]/[55(50)] lb Cost/lb = $40(50) / [55(50)] lb is the correct answer.

Modeling Using Equations or Inequalities (7.EE.B.3)

1	D	The first step to solving this problem is to figure out the value of t, the number of hours. As you know, 7 am to 3 pm represents 5 hours to 12 pm, then another 3 hours to 3pm, for a total of 8 hours. This gives us W = 3.75(8) = 30, a full tank.
2	D	Let the volume of water = w and volume of milk = m. m 3 = w 8 Rearranging the above equation, m = 3w/8 m + w = 187 cubic ft. Replacing the value of m by 3w/8 in the above equation, 3w/8 + w = 187 11w/8 = 187 w = (187*8)/11 w = 136 cubic ft.
3	D	Remember: The volume of a rectangular box is V = lwh. This lets us set up the following equation: 960 = 12(10)h 960 = 120h 8 = h This gives us a height of 8 inches, as indicated.
4	D	Let b = the number of bushels harvested 24/12 = 18/b 24b = 12(18) 24b = 216 b = 9 bushels is the correct answer.
5	C	Let m = Melanie's age Then 27-m = Roni's Age m - (27-m) = 3 m - 27 + m = 3 2m - 27 = 3 2m = 3 + 27 2m = 30 m = 15 27 - m = 27 - 15 27 - m = 3 Melanie's Age = 15, and Roni's Age = 12 is the correct answer.
6	B	Solving this problem requires the creation of a simple equation: 2.5(x) = 50 Dividing 50 by 2.5, the answer is 20.

Question No.	Answer	Detailed Explanation
7	B	$11 + (8 + 6) = (y + 8) + 6$ This equation demonstrates the associative property for addition, which states that when adding three or more quantities, the way that they are grouped does not affect the sum. Therefore, $y = 11$ because of the associative property of addition.
8	C	$p = 2c + 1$ $p = 2(7) + 1$ $p = 14 + 1$ $p = 15$ people present 15 is the correct answer.
9	A	Let the volume of salt = x and volume of sugar = y x 4 = y 8 Rearranging the above equation, x = 4w/8 m + w = 300 cubic ft. Replacing the value of m by 4w/8 in the above equation, 4w/8 + w = 300 12w/8 = 300 w = (300*8)/12 w = 200 cubic ft.
10	A	Apply the rule to each number's position in the sequence: (1) $(8 + 1^2) - 2 \times 1 = 7$ (2) $(8 + 2^2) - 2 \times 2 = 8$ (3) $(8 + 3^2) - 2 \times 3 = 11$ (4) $(8 + 4^2) - 2 \times 4 = 16$ (5) $(8 + 5^2) - 2 \times 5 = 23$ Therefore, sequence 7, 8, 11, 16, 23, ... follows the rule $(8 + t^2) - 2t$.

Solving Multi-Step Problems (7.EE.B.4.A)

1	A	In order to find out how much Bill should charge his client, divide 130 by 4, and then add the quotient to 130: (1) $130 \div 4 = 32.50$ (2) $130 + 32.50 = \$162.50$
2	C	First, write the expressions based on the language in the problem: length = 6x width = (1/3)(6x) height = (1/2)(6x) Next, solve for x based on the formula for volume, lwh. (1) $6x \times ((1/3)(6x)) \times ((1/2)(6x)) = 972$ (2) $6x \times 2x \times 3x = 972$ (3) $12x^2 \times 3x = 972$ (4) $36x^3 = 972$ (divide each side by 36) (5) $x^3 = 27$ (6) $x = 3$
3	D	$x + 2x + x + 2x = 180$ ft. $6x = 180$ ft. $x = 30$ ft. $2x = 60$ ft. $x + 2x + x = 120$ ft is the correct answer.
4	D	Now: In 2 years: Jim = x + 4 x + 6 Bob = x x + 2 x + 6 = 2(x + 2) x + 6 = 2x + 4 2 = x = Bob 6 = x + 4 = Jim Bob is 2, and Jim is 6.
5	C	Let x = number of girls 2x = number of boys x + 2x = 39 3x = 39 x = 13 girls 2x = 26 boys
6	A	First, set up the equation based on the language of the problem: (1) Amount of gas = 3 gallons (2) Gas level = 10% (3) Tank capacity = 12 gallons (4) x = Amount of gallons needed to fill tank (5) Equation = 12 - (12(0.10) + 3) = x Next, solve for x (6) 12 - (1.2 + 3) = x (7) 12 - 4.2 = x (8) 7.8 gallons

Question No.	Answer	Detailed Explanation
7	C	Let n = number of weeks of work for Sue to catch up with Melanie 35 + 15n = 20n 35 = 5n 7 = n 7 weeks is the correct answer.
8	D	6 dozen = 72 cookies 128 / 72 = 1 56/72 = 1 7/9 recipes 1 7/9 x 3 1/2 = 6.22 cups 6 c is the correct answer.
9	B	She spent 4 x $4.00 = $16.00 for blueberries and 5 x $1.49 = $7.45 for apples and $6.00 for potatoes and $8.00 for a pie $37.45 Total $40.00 - $37.45 = $2.55 left $2.55/$40.00 =.06375= 6.375% left
10	B	8/8 - 7/8 = 1/8 on lawnmower (1/8)s = 250 s = 8(250) s = $2000 (total savings) 7/8 (2000) = 14000/8 = 1750 $1750 is the correct answer.

Linear inequality word problems (7.EE.B.4.B)

1	C	3000 must be multiplied by y, the years of experience. This amount must then be added to the base salary of $50,000. This sum must be less than or equal to $70,000.
2	B	The weekly increase of 45 is multiplied by the number of weeks. This amount is added to the initial savings of 225. This sum must be greater than or equal to 500.
3	D	The number of minutes of charging time must be multiplied by 2 and divided by 3 to find the percentage of the battery that is charged over that time. This is then added to the initial charge of 20 percent, and the sum must be at least 75.
4	A	Set up an inequality that $74 + 0.20t \geq 90$. Solving for t, we find that $t \geq 80$.
5	C	Set up an inequality that $100 - 4n \geq 83$. Solving for n, we find that $n \leq 4.25$. Because you can only miss whole numbers of problems, the most he can miss is 4.
6	B	In the inequaliy, let T be 12, and solve for L. The solution becomes that L is less than or equal to 45.
7	C	If the number of units sold is 0, the 300 is still present as the minimum amount of the pay. The 25 is multiplied by 0 in the equation and thus negated, but the 300 remains.
8	B	3.75 hours is just enough for750 - 200h to negate the 750 Calories he would otherwise gain in a given week.
9	B	Because there are twice as many girls as boys to be admitted, girls must make up 2/3 of the adimtted group. This leaves one third to be boys. One third of 300 is 100.
10	A	His minimum hours of chores increased by 1.5 hours. This must represent 6 hours of TV watching that week .

Geometry

Scale Models (7.G.A.1)

1. **Triangle ABC and triangle PQR are similar. Find the value of x.**

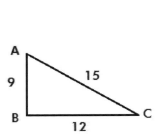 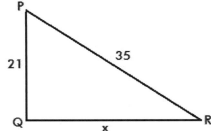

   Ⓐ 23
   Ⓑ 25
   Ⓒ 26
   Ⓓ 28

2. **If the sides of two similar figures have a similarity ratio of $\dfrac{3}{2}$ what is the ratio of their areas?**

   Ⓐ $\dfrac{9}{4}$

   Ⓑ $\dfrac{3}{2}$

   Ⓒ $\dfrac{1}{3}$

   Ⓓ $\dfrac{3}{1}$

3. **What is the similarity ratio between the following two similar figures?**

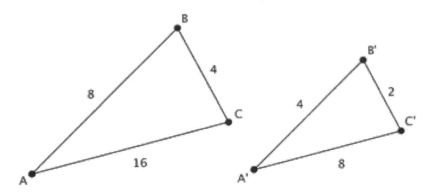

Ⓐ $\dfrac{2}{1}$

Ⓑ $\dfrac{1}{4}$

Ⓒ $\dfrac{2}{3}$

Ⓓ $\dfrac{3}{3}$

4. **A map is designed with a scale of 1 inch for every 5 miles. If the distance between two towns is 3 inches on the map, how far apart are they?**

Ⓐ 15 miles
Ⓑ 3 miles
Ⓒ 5 miles
Ⓓ 1.5 miles

5. **If the angles of one of two similar triangles are 30, 60, and 90 degrees, what are the angles for the other triangle?**

Ⓐ 60, 120, 180
Ⓑ 45, 45, 90
Ⓒ 30, 60, 90
Ⓓ There is not enough information to determine.

6. The ratio of similarity between two figures is $\frac{2}{3}$.

    If one side of the larger figure is 12 cm long, what is the length of the corresponding side in the smaller figure?

    Ⓐ 9 cm
    Ⓑ 16 cm
    Ⓒ 3 cm

    Ⓓ $\frac{4}{3}$ cm

7. If the sides of two similar figures have a similarity ratio of $\frac{5}{3}$ what is the ratio of their perimeters?

    Ⓐ $\frac{25}{9}$

    Ⓑ $\frac{5}{6}$

    Ⓒ $\frac{5}{3}$

    Ⓓ $\frac{3}{5}$

8. Triangle ABC and triangle DEF are similar. Find the value of x.

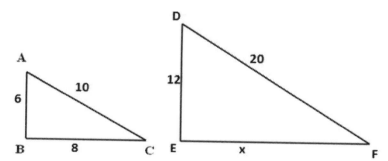

    Ⓐ 16
    Ⓑ 18
    Ⓒ 14
    Ⓓ 12

9. **Triangle ABC and triangle DEF are similar. Find the value of x.**

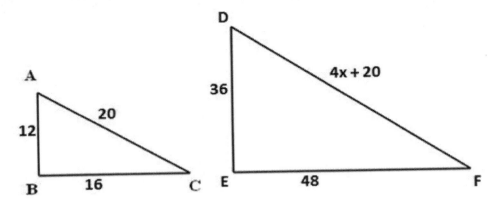

- (A) 5
- (B) 20
- (C) 10
- (D) 15

10. **Triangle ABC and triangle DEF are similar. Find the value of x.**

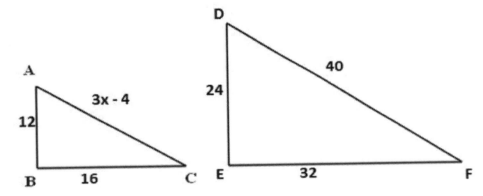

- (A) 5
- (B) 10
- (C) 8
- (D) 6

Drawing Plane (2-D) Figures (7.G.A.2)

1. Which of the following lengths cannot be the lengths of the sides of a triangle?

   Ⓐ 4, 6, 9
   Ⓑ 3, 4, 2
   Ⓒ 2, 2, 3
   Ⓓ 1, 1, 2

2. Which of the following set of lengths cannot be the lengths of the sides of a triangle?

   Ⓐ 12.5, 20, 30
   Ⓑ 10, 10, 12
   Ⓒ 4, 8.5, 14
   Ⓓ 3, 3, 3

3. If the measure of two angles in a triangle are 60 and 100 degrees, what is the measure of the third angle?

   Ⓐ 20 degrees
   Ⓑ 50 degrees
   Ⓒ 30 degrees
   Ⓓ 180 degrees

4. Which of the following triangle classifications does not describe the angles in a triangle?

   Ⓐ Right
   Ⓑ Acute
   Ⓒ Equiangular
   Ⓓ Scalene

5. Which of the angles has the least measure?

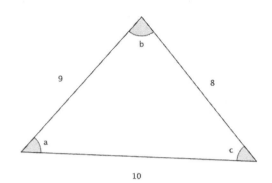

- Ⓐ a
- Ⓑ b
- Ⓒ c
- Ⓓ There is not enough information to tell

6. Find the value of x.

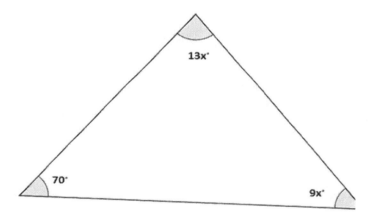

- Ⓐ 6
- Ⓑ 5
- Ⓒ 10
- Ⓓ 4

7. Which of the following lengths cannot be the lengths of the sides of a triangle?

- Ⓐ 8, 12, 18
- Ⓑ 6, 8, 4
- Ⓒ 4, 4, 6
- Ⓓ 2, 2, 4

8. If this is an isosceles triangle, which of the following could be the measure of each of the unknown angles?

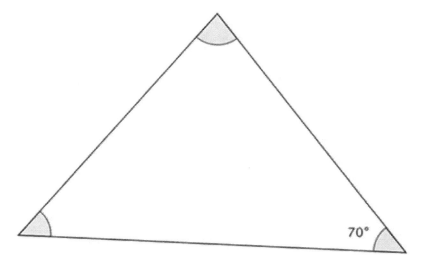

70°

  Ⓐ 55°
  Ⓑ 66°
  Ⓒ 53°
  Ⓓ 70°

9. Which of the following triangle classification is defined by an angle that is created by two perpendicular lines?

  Ⓐ Right
  Ⓑ Acute
  Ⓒ Equiangular
  Ⓓ Scalene

10. Which of the following triangle classifications is defined by three angles less than 90 degrees?

  Ⓐ Right
  Ⓑ Acute
  Ⓒ Obtuse
  Ⓓ Scalene

Cross Sections of 3-D Figures (7.G.A.3)

1. **The horizontal cross section of a cylinder is a _____.**

   Ⓐ rectangle
   Ⓑ triangle
   Ⓒ circle
   Ⓓ parallelogram

2. **In order for a three-dimensional shape to be classified as a "prism," its horizontal cross-sections must be _____.**

   Ⓐ congruent polygons
   Ⓑ non-congruent polygons
   Ⓒ circles
   Ⓓ equilateral triangles

3. **Which of the following nets is not the net of a cube?**

   Ⓐ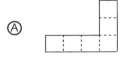

   Ⓑ

   Ⓒ

   Ⓓ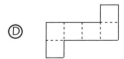

4. **If you were to "unwrap" a cylinder, what shape represents the side of the cylinder?**

   Ⓐ Circle
   Ⓑ Triangle
   Ⓒ Trapezoid
   Ⓓ Rectangle

© Lumos Information Services 2015 | LumosLearning.com

5. **The vertical cross section of a cylinder with its circular bases at the top and bottom is a _____.**

   Ⓐ Rectangle
   Ⓑ Triangle
   Ⓒ Circle
   Ⓓ Semicircle

6. **Which of the following shapes represents the sides of a pyramid?**

   Ⓐ Triangle
   Ⓑ Rectangle
   Ⓒ Trapezoid
   Ⓓ Square

7. **How many faces does a cube have?**

   Ⓐ 4
   Ⓑ 2
   Ⓒ 6
   Ⓓ 8

8. **The cross section of a sphere is a _____.**

   Ⓐ Semicircle
   Ⓑ Triangle
   Ⓒ Circle
   Ⓓ Parallelogram

9. **Which statement is not true?**

   Ⓐ The right cross section of a square pyramid is a trapezoid.
   Ⓑ The vertical cross section of a square pyramid is a triangle.
   Ⓒ The base of a square pyramid is a triangle.
   Ⓓ The horizontal cross section of a square pyramid is a square.

10. **Samuel and Christina made a model colonial-style house for their history class. The figures below show the top, side, and front views of the two 3D figures they used to make their house.**

    Based on the figures, which two 3D figures did Samuel and Christina use?

    Ⓐ Triangular prism and cube
    Ⓑ Rectangular prism and triangular prism
    Ⓒ Triangular pyramid and triangular prism
    Ⓓ Cube and triangular pyramid

Solve real-life and mathematical problems involving angle measure, area, surface area, and volume

Circles (7.G.B.4)

1. A circle is divided into 4 equal sections. What is the measure of each of the angles formed at the center of the circle?

   Ⓐ 25°
   Ⓑ 180°
   Ⓒ 90°
   Ⓓ 360°

2. What is the area of a circle with diameter 8 cm? Round your answer to the nearest tenth.

   Ⓐ 201.1 cm²
   Ⓑ 201.0 cm²
   Ⓒ 50.2 cm²
   Ⓓ 25.1 cm²

3. What is the radius of a circle with a circumference of 125 cm? Round your answer to the nearest whole number.

   Ⓐ 24 cm
   Ⓑ 10 cm
   Ⓒ 20 cm
   Ⓓ 19 cm

4. What is the circumference of a circle with radius 0.5 feet? Round your answer to the nearest tenth.

   Ⓐ 3.1 ft
   Ⓑ 3.2 ft
   Ⓒ 0.8 ft
   Ⓓ 0.7 ft

5. Which of the following could constitute the area of a circle?

   Ⓐ 50 units
   Ⓑ 1 square unit
   Ⓒ 1.5 cubic units
   Ⓓ One half of a unit

6. **If two radii form a 30 degree angle at the center of a circle with radius 10 cm, what is the area between them? Round your answer to the nearest tenth.**

   Hint: A circle "sweeps out" 360 degrees.

   Ⓐ 26.2 square centimeters
   Ⓑ 26.1 square centimeters
   Ⓒ 314.1 square centimeters
   Ⓓ 314.2 square centimeters

7. **What is the area of a circle with radius 2.8 cm? Round your answer to the nearest tenth.**

   Ⓐ 24.7 cm²
   Ⓑ 24.6 cm²
   Ⓒ 17.6 cm²
   Ⓓ 8.8 cm²

8. **What is the radius of a circle with area 50 square cm? Round your answer to the nearest whole number.**

   Ⓐ 4 cm
   Ⓑ 3 cm
   Ⓒ 8 cm
   Ⓓ 7 cm

9. **What is the perimeter of a semi-circle with radius 8 in? Round your answer to the nearest tenth.**

   Hint: A semi-circle is half a circle.

   Ⓐ 201.0 in
   Ⓑ 50.2 in
   Ⓒ 100.5 in
   Ⓓ 41.1 in

10. **What is the diameter of a circle with an area of 50.24 m²?**

   Ⓐ 4 m
   Ⓑ 6 m
   Ⓒ 8 m
   Ⓓ 16 m

Angles (7.G.B.5)

1. **Find x.**

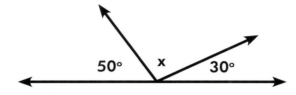

   Ⓐ 40°
   Ⓑ 60°
   Ⓒ 80°
   Ⓓ 100°

2. **Find the measures of the missing angles in the figure below.**

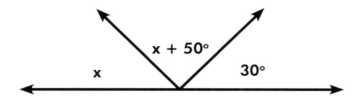

   Ⓐ 30° and 60°
   Ⓑ 60° and 90°
   Ⓒ 50° and 100°
   Ⓓ 60° and 120°

3. **The sum of the measures of angles a and b 155 degrees. What is the measure of angle b?**

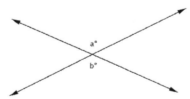

   Ⓐ 155 degrees
   Ⓑ 77.5 degrees
   Ⓒ 35 degrees
   Ⓓ 210.5 degrees

4. **What is true about every pair of vertical angles?**

Ⓐ They are supplementary.
Ⓑ They are complementary.
Ⓒ They are equal in measure.
Ⓓ They total 360 degrees.

5. **If the measures of two angles have a sum of 180 degrees, they are called —**

Ⓐ supplementary angles
Ⓑ complementary angles
Ⓒ vertical angles
Ⓓ equivalent angles

6. **If angle a measures 30 degrees, what is the measure of angle b?**

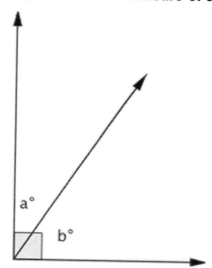

Ⓐ 60 degrees
Ⓑ 30 degrees
Ⓒ 150 degrees
Ⓓ 20 degrees

7. **If the measure of the first of two complementary angles is 68 degrees, what is the measure of the second angle?**

Ⓐ 68 degrees
Ⓑ 22 degrees
Ⓒ 44 degrees
Ⓓ 34 degrees

8. **If the sum of the measures of angles a and b is 110 degrees, what is the measure of angle c?**

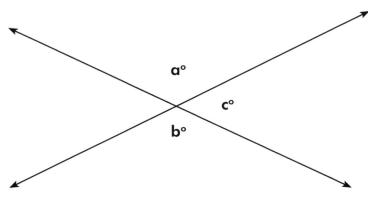

  Ⓐ 125 degrees
  Ⓑ 55 degrees
  Ⓒ 70 degrees
  Ⓓ 180 degrees

9. **If two angles are both supplementary and equal in measure, they must be**

  Ⓐ vertical angles
  Ⓑ right angles
  Ⓒ adjacent angles
  Ⓓ obtuse angles

10. **If the sum of the measures of angles a and b is 240 degrees, what is the measure of angle c?**

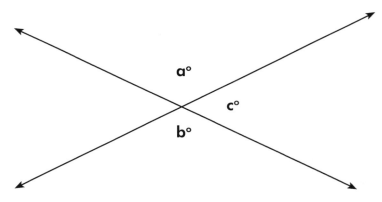

  Ⓐ 60 degrees
  Ⓑ 30 degrees
  Ⓒ 160 degrees
  Ⓓ 150 degrees

Finding Area, Volume, & Surface Area (7.G.B.6)

1. **Find the area of the rectangle shown below.**

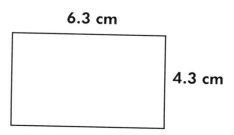

6.3 cm

4.3 cm

- Ⓐ 10.5 square centimeters
- Ⓑ 24 square centimeters
- Ⓒ 27.09 square centimeters
- Ⓓ 21 square centimeters

2. **What is the volume of a cube whose sides measure 8 inches?**

- Ⓐ 24 in³
- Ⓑ 64 in³
- Ⓒ 128 in³
- Ⓓ 512 in³

3. **Calculate the area of the following polygon.**

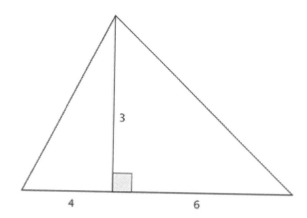

3

4   6

- Ⓐ 15 square units
- Ⓑ 30 square units
- Ⓒ 36 square units
- Ⓓ 18 square units

4. **Calculate the area of the following polygon.**

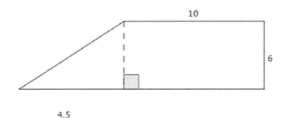

Ⓐ 60 square units
Ⓑ 73.5 square units
Ⓒ 13.5 square units
Ⓓ 24 square units

5. **What is the volume of the following triangular prism?**

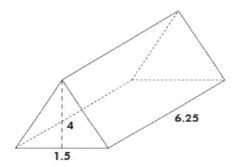

Ⓐ 11.75 cubic units
Ⓑ 20 cubic units
Ⓒ 37.5 cubic units
Ⓓ 18.75 cubic units

6. **What is the volume of a prism with the following base and a height of 2.75?**

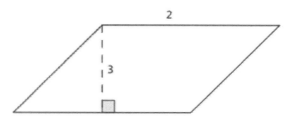

Ⓐ 8.25 cubic units
Ⓑ 13.75 cubic units
Ⓒ 16.5 cubic units
Ⓓ 8.75 cubic units

7. What is the surface area of a cube with sides of length 2?

- Ⓐ 16 square units
- Ⓑ 8 square units
- Ⓒ 24 square units
- Ⓓ 18 square units

8. What is the surface area of a rectangular prism with dimensions 2, $\dfrac{1}{2}$, and $\dfrac{1}{4}$?

- Ⓐ 2

- Ⓑ $\dfrac{13}{4}$

- Ⓒ $\dfrac{9}{4}$

- Ⓓ $\dfrac{3}{2}$

9. Find the area of the shape below. (Round to the nearest tenth). Use pi = 3.14.

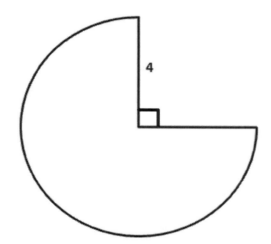

- Ⓐ 37.7 square units
- Ⓑ 50.2 square units
- Ⓒ 18.8 square units
- Ⓓ 35.2 square units

10. John has a container with a volume of 170 cubic feet filled with sand. He wants to transfer his sand into the new container shown below so he can store more sand. After he transfers the sand, how much more sand can he add to the new container?

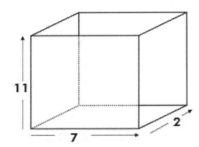

   Ⓐ 16 cubic feet of sand
   Ⓑ 26 cubic feet of sand
   Ⓒ 150 cubic feet of sand
   Ⓓ 324 cubic feet of sand

End of Geometry

Geometry

Answer Key
&
Detailed Explanations

Scale Models (7.G.A.1)

Question No.	Answer	Detailed Explanation
1	D	Remember: In order to solve a similarity question, set up a proportion with corresponding sides, and solve!: x/12 = 35/15 or x/12 = 21/9 You can use cross products (ad = bc) to solve for x. (1) x/12 = 35/15 (2) 15x = (35)(12) (3) 15x = 420 (4) x = 420 ÷ 15 (5) x = 28
2	A	If the similarity ratio is 3/2, then the ratio of the areas is the square of that ratio: 3/2 × 3/2 = 9/4.
3	A	In order to solve a similarity problem, set up a proportion with corresponding sides: 8/4 = 16/8. Both ratios in simplest form are 2/1. Therefore, the similarity ratio is 2/1.
4	A	You can use cross products (ad = bc) to solve for x. (inch)/(miles) = (inch)/(miles) (1) 1/5 = 3/x (2) 1x = (3)(5) (3) x = 15 Therefore, the towns are 15 miles apart.
5	C	Corresponding angles are congruent in similar figures. Thus, the angles will have the same measures in the second figure as in the first.
6	A	In order to find the length of the smaller side, divide 12 by 4/3. (1) 12/1 ÷ 4/3 (2) 12/1 × 3/4 (3) (12 × 3)/(1 × 4) (4) 36/4 (5) 9. Therefore, the length of the corresponding side is 9 cm.
7	C	In order to solve a similarity problem, set up a proportion with corresponding sides. If the similarity ratio is 5/3, then the ratio of the perimeters is 5/3.
8	A	You can use cross products (ad = bc) to solve for x. (1) 10/20 = 8/x (2) 10x = (20)(8) (3) 10x = 160 (divide both sides by 10) (4) x = 16
9	C	You can use cross products (ad = bc) to solve for x. (1) 16/48 = 20/(4x + 20) (2) (16)(4x + 20) = (48)(20) (3) 64x + 320 = 960 (subtract both sides by 320) (4) 64x = 640 (divide each side by 64) (5) x = 10
10	C	You can use cross products (ad = bc) to solve for x. (1) 16/32 = (3x - 4)/(40) (2) (16)(40) = (32)(3x - 4) (3) 640 = 96x - 128 (add both sides by 128) (4) 768 = 96x (divide each side by 96) (5) 8 = x

Question No.	Answer	Detailed Explanation
		## Drawing Plane (2-D) Figures (7.G.A.2)
1	D	The sum of any two sides of a triangle is always greater than the third side. Here, the sum of $1 + 1 = 2$. Since the sum of $1 + 1$ is not greater than 2, the lengths given cannot be the side lengths of a triangle.
2	C	The sum of any two sides of a triangle is always greater than the third side. Here, the sum of $4 + 8.5 = 12.5$. Since the sum of $4 + 8.5$ is not greater than 14, the lengths given cannot be the side lengths of a triangle.
3	A	The sum of the measures of the angles in a triangle is always $180°$. To find a missing angle, add the known sides, and subtract the sum from $180°$. (1) $100 + 60 = 160$ (2) $180 - 160 =$ (3) 20 Therefore, the measure of the third angle is $20°$.
4	D	Triangles classified by angles are acute (all acute angles), obtuse (one obtuse angle), or right (one right angle). Triangles classified by sides are scalene (no equal sides), isosceles (two equal sides) or equilateral (three equal sides).
5	A	The answer is a. In triangles, the smallest angle (by degree measure) is always across from the shortest side. Since 8 is the length of the shortest side, a is the smallest angle because it is across from 8.
6	B	The sum of the measures of the angles in a triangle is always $180°$. To find a missing angle, add the known sides, and solve for x. (1) $70 + 13x + 9x = 180$ (2) $70 + 22x = 180$ (now subtract 70 from both sides) (3) $22x = 110$ (divide each side by 22) (4) $x = 5$
7	D	The sum of any two sides of a triangle is always greater than the third side. Here, the sum of $2 + 2 = 4$. Since the sum of $2 + 2$ is not greater than 4, the lengths given cannot be the side lengths of a triangle.
8	A	The answer is $55°$. The sum of $55 + 55$ is 110. 110 plus the known angle equals $180°$. Therefore, the measure of each missing angle is $55°$.
9	A	When two lines intersect to form perpendicular lines, four $90°$ angles are created. Since $90°$ angles are also called right angles, a triangle that has one angle formed by perpendicular lines is a right triangle.

Question No.	Answer	Detailed Explanation
10	B	Triangles classified by angles are acute (all acute angles, which are angles less than 90°), obtuse (one obtuse angle), or right (one right angle). Triangles classified by sides are scalene (no equal sides), isosceles (two equal sides), or equilateral (three equal sides).

Cross Sections of 3-D Figures (7.G.A.3)

Question No.	Answer	Detailed Explanation
1	C	A cylinder has two circular bases and one curved surface that appears as a rectangle when the cylinder is shown as a net. Therefore, a horizontal cross section would show a circle.
2	A	A prism consists of two congruent bases and various congruent faces; therefore, the cross sections of a prism must show congruent polygons.
3	A	In order to form a cube, the nets must fold together to make the shape. Here, the figure would not form a cube when folded together to make a three-dimensional figure.
4	D	A cylinder has two circular bases and one curved surface that appears as a quadrilateral (rectangle) when the cylinder is shown as a net. Therefore, a side of a cylinder would show a rectangle.
5	A	A cylinder has two circular bases and one curved surface that appears as a quadrilateral (rectangle) when the cylinder is shown as a net. Therefore, a vertical cross section of a cylinder would show a rectangle.
6	A	A pyramid consists of 1 polygon base and all triangular lateral faces; therefore, the side of pyramid would show a triangle.
7	C	A cube is a rectangular prism with four lateral faces and two bases (also faces) which are square. Therefore, the total number of faces for a cube is 6.
8	C	A sphere is a shape with a curved surface. A cross section of a sphere is a circle.
9	C	The base of a square pyramid is a square, not a triangle.
10	B	The first 2D figure represents a 3D shape with 6 rectangular faces. The second 2D figure represents a 3D shape with two triangular bases and 3 rectangular faces. Therefore, Samuel and Christina used a rectangular prism and a triangular prism for their house.

Question No.	Answer	Detailed Explanation

Circles (7.G.B.4)

1	C	The correct answer is 90°. Selecting 25° results from incorrectly applying the qualities of a circle graph to the circle (a whole circle represents 100%, so four equal parts equal 25% each). Choosing 180° is a result of measuring the angle lengths of the lines formed by dividing the circle into four equal parts (a straight line measures 180°). A circle measures 360°, so the measure of each angle formed at the center would be less than 360°. By dividing it into 4 equal pieces, each angle will be 360 ÷ 4 = 90°.
2	C	The correct answer is 50.2 cm². To find the area of a circle, apply the formula $A = \pi r^2$. Since the problem gives the diameter of the circle, the first step is to find the radius by dividing the diameter by 2. 8 ÷ 2 = 4 cm. Next, plug in the numbers into the formula: (1) $\pi 4^2$ = (2) 3.14 x 4² = (3) 3.14 x 16 = (4) 50.24 cm² (5) 50.2 cm² (rounded to the nearest tenth). Common errors made when applying the area formula to circles are multiplying the radius by 2 instead of by itself (which would result in 25.1 cm²) or using the diameter of the circle to find the area (resulting in 201.0 cm²).
3	C	The correct answer is 20 cm. To find the radius when given the circumference of the circle, use C/(2π), where C equals circumference. Insert the numbers from the problem, and solve: (1) r = C/(2π) (2) 125 ÷ (2 x 3.14) = (3) 125 ÷ 6.28 = (4) 19.90 cm = (5) 20 cm (rounded to the nearest whole number). Finding a radius of 19 cm results from rounding down instead of rounding up. An answer of 10 cm results from dividing the radius by 2. Choosing 24 cm is based on adding 2 + π and dividing 125 by the sum.
4	A	The answer is 3.1 ft. Finding 3.2 ft results from rounding up instead of rounding down. A result of 0.8 ft comes from applying the formula for the area of a circle. Selecting 0.7 ft also results from using the area formula but also includes a rounding error. To apply the formula for the circumference of a circle, 2πr, plug in the given values, and solve. (1) 2πr (2) 2 × 3.14 × 0.5 (3) 6.28 × 0.5 = (4) 3.14 ft = (5) 3.1 ft (rounded to the nearest tenth).

Question No.	Answer	Detailed Explanation
5	B	The answer is 1 square unit. Area is defined as the number of square units that cover a specific space. Since the question calls for the area of a circle, the best answer is 1 square unit.
6	A	The answer is 26.2 square centimeters. First, find the area of the entire circle by applying the area formula—πr^2. Next, calculate the number of square centimeters per degree by dividing the area of the circle by the number of degrees in the circle. Multiply the quotient by 30 and round to the nearest tenth. (1) πr^2 (2) $3.14 \times 10^2 =$ (3) $3.14 \times 100 =$ (4) Area = 314 cm² (5) $314 \div 360 = 0.87$ per degree (6) $30 \times 0.87 = 26.16$ square centimeters (7) 26.2 cm² (rounded to the nearest tenth)
7	B	The correct answer is 24.6 cm². Finding 17.6 cm² results from first multiplying 2.8×2 and then multiplying the product by 3.14. An area of 8.8 cm² comes from multiplying 3.14×2.8. To apply the formula for the area of a circle, πr^2, plug in the given values, and solve. (1) πr^2 (2) $3.14 \times 2.8^2 =$ (3) $3.14 \times 7.84 =$ (4) 24.61 cm² = (5) 24.6 cm² (rounded to the nearest tenth)
8	A	The answer is 4 cm. To apply the formula for finding the radius when given the area of a circle, which is the square root of A/π, plug in the given values and solve. (1) r = square root of A/π (2) square root of $(50 \div 3.14) =$ (3) square root $(15.92) =$ (4) 3.99 cm = (5) 4 cm (rounded to the nearest whole number). Finding 3 cm results from rounding down instead of rounding up. A result of 8 cm comes from dividing the quotient by 2 instead of finding the square root. A radius of 7 cm results from dividing by 2 and rounding down instead of rounding up.
9	D	The answer is 41.1 in. Perimeter = 1/2 circle = diamter $1/2 (2\pi8) = (8 + 8)$ $8\pi + 16$ $8 (3.14) +16$ $25.1 + 16$ 41.1

Question No.	Answer	Detailed Explanation
10	C	An answer of 4 m results from calculating the radius by dividing 50.24 ÷ 3.14 and then finding the square root of the quotient, 16. Selecting 6 m results from following the preceding steps, but then adding 4 + 2. Choosing 16 m is the result of dividing 50.24 ÷ 3.14. In order to find the diameter, apply the formula for diameter after finding the radius. The formula is d = 2r (diameter = 2 × radius). As mentioned, the radius is 4 m. Therefore, 4 × 2 = 8 m.

Angles (7.G.B.5)

Question No.	Answer	Detailed Explanation
1	D	The answer is 100°. Reminder: Angles that together form a straight line are called supplementary, meaning they add to 180 degrees. In this case, 50 + x + 30 = 180 requires an x value of 100 degrees.
2	C	The answer is 50° and 100°. Angles that together form a straight line are supplementary, meaning their measures add to 180 degrees. In this case, x + (x + 50) + 30 = 180 can only be satisfied by an x value of 50, resulting in angles of measure 50 and 100 degrees.
3	B	The answer is 77.5 degrees. When two lines intersect, they form vertical angles, which are equal in measure. Angle a and angle b are vertical angles. To find the value of angle b, divide 155 by 2. The quotient is the value of angle b: 155 ÷ 2 = 77.5 degrees.
4	C	Vertical angles are congruent angles formed by two intersecting lines. The sum of the angles can be less than or greater than 90 degrees and 180 degrees, respectively, so they are not necessarily complementary or supplementary angles. Also, vertical angles do not form a complete circle, so they do not total 360 degrees.
5	A	Two angles are supplementary if the sum of their measures is equal to 180°. The sum of the measures of the angles cannot be greater than or less than 180°. It must be exactly 180°.
6	A	The answer is 60°. Angles a and b form a right angle, which measures 90°. This means that angles a and b are complementary. To find the measure of angle b, subtract the measure of angle a from 90: 90 - 30 = 60. Therefore, the measure of angle b is 60°.

Question No.	Answer	Detailed Explanation
7	B	The answer is 22°. Two angles are complementary if the sum of the measures of the angles equals 90°. To find the measure of the second angle, subtract the measure of the first angle from 90: 90 - 68 = 22. Therefore, the measure of the second angle is 22°.
8	A	The answer is 125 degrees. When two lines intersect to form vertical angles, the opposite angles are equal in measure. Angle a and angle b are vertical angles. To find the value of angle a, divide 110 by 2. The quotient is the value of angle a: 110 ÷ 2 = 55 degrees. Two intersecting lines also form adjacent supplementary angles which add up to 180°. Since angle a and angle c are adjacent and supplementary, subtract 55 from 180 to find the measure of angle c: 180 - 55 = 125.
9	B	Two angles are supplementary if the sum of the measures of the angles equals 180°. Since right angles are 90°, the sum of two right angles is 180°. Therefore, two right angles form supplementary angles.
10	A	The answer is 60 degrees. When two lines intersect to form vertical angles, the opposite angles are equal in measure. Angle a and angle b are vertical angles. To find the value of angle a and angle b, divide 240 by 2: 240 ÷ 2 = 120 degrees. Two intersecting lines also form adjacent supplementary angles, which add up to 180 degrees. Since angle a and angle c are adjacent and supplementary, subtract 120 from 180 to find the measure of angle c: 180 - 120 = 60.

Finding Area, Volume, & Surface Area (7.G.B.6)

1	C	The answer is 27.09 square centimeters. To calculate the area of a rectangle, multiply the length and width. Multiplying $6.3 \times 4.3 = 27.09$ square centimeters. An answer of 10.5 square centimeters results from adding $6.2 + 4.3$. Choosing 24 square centimeters is the result of rounding 6.2 and 4.3 to 6 and 4 before multiplying. Selecting 21 square centimeters results from adding $6.2 + 6.2 + 10.5 + 10.5$.
2	D	The answer is 512 in^3. The formula for the volume for a cube is $V = s3$, where s is the length of one side. Multiplying $8 \times 8 \times 8 = 512 \text{ in}^3$. An answer of 24 in^3 results from adding $8 + 8 + 8$. Finding 64 in^3 is the result of using the area formula, s^2. Selecting 128 in^3 results from multiplying $8 \times 8 \times 2$.

Question No.	Answer	Detailed Explanation
3	A	The correct answer is 15 square units. To find the correct answer, apply the formula for the area of a triangle, A = ½bh. First, calculate the base by adding 6 + 4 = 10. Next, multiply the base times the height: 10 × 3 = 30. Then, divide the product by 2: 30 ÷ 2 = 15 square units. Finding an answer of 30 square units results from multiplying the base times the height and not dividing the product by 2. Selecting 36 square units is the result of multiplying 4 × 6 to find the base, multiplying the base, 24, by the height, 3, and dividing the product by 2: 72 ÷ 2 = 36 Choosing 18 square units results from using 6 as the base, multiplying the base by 3, and not dividing the product by 2.
4	B	The answer is 73.5 square units. This is a compound figure comprised of a triangle and a rectangle. Identify the length and width of each rectangle. Apply the formula for area of a rectangle, A = bh or A = lw, and the formula for the area of a triangle, A = ½bh. Then, add the two products together. The sum is the total area of the figure. (1) area of rectangle one: 6 × 10 = 60 square units (2) area of rectangle two: 4.5 × 6 = 27 ÷ 2 = 13.5 square units (3) 60 + 13.5 = 73.5 square units
5	D	The answer is 18.75 cubic units. The formula for volume is V = Bh, where B = the area of the base, and h = height. Since the base is a triangle, the formula is V = (½bh)(h). To solve, plug in the numbers: (1) (½bh)(h) = (2) (½ × 1.5 × 4)(6.25) = (3) (3)(6.25) = (4) 18.75 cubic units
6	C	The answer is 16.5 cubic units. The formula for volume is V = Bh, where B = the area of the base, and h = height. Since the base is a parallelogram, the formula is V = (bh)(h). To solve, plug in the numbers: (1) (bh)(h) = (2) (2 × 3)(2.75) = (3) (6)(2.75) = (4) 16.5 cubic units
7	C	A cube has 6 square faces. In this particular cube, each face has an area of 2 x 2 = 4 square units. The overall surface area = 6 x 4 = 24 square units
8	B	This prism is made up of 6 rectangles. Two of them are 2 by 0.5, two of them are 2 by 0.25, and two of them are 0.5 by 0.25 The surface area = 2(2)(0.5) + 2(2)(0.25) + 2(0.5)(0.25) = 3.25 square units, or 13/4 square units

Question No.	Answer	Detailed Explanation
9	A	The answer is 37.7 square units. First, find the area of the entire circle by applying the area formula—πr^2. Next, multiply the area by 3/4 in order to find the area of the shape. (1) πr^2 (2) $3.14 \times 4^2 =$ (3) $3.14 \times 16 =$ (4) Area $= 50.24$ cm^2 (5) $50.24 \times 3/4 = 37.68 =$ (6) $37.68 = 37.7$ cm^2
10	A	The answer is 16 cubic feet of sand. John can add 16 cubic feet of sand after filling the new container. Find the volume of the new container by multiplying $2 \times 7 \times 11$. The product is 154 cubic feet. Subtract 154 from 170. The difference is 16 cubic feet. The answer of 26 cubic feet results from not regrouping when subtracting 170 - 154. Finding an answer of 150 cubic feet is a result of adding $2 + 7 + 11$ and subtracting the sum of 20 from 170. Choosing 324 cubic feet results from adding $170 + 154$.

Statistics and Probability

1. Joe and Mary want to calculate the average height of students in their school. Which of the following groups of students would produce the least amount of bias?

   Ⓐ Every student in the 8th grade.
   Ⓑ Every student on the school basketball team.
   Ⓒ A randomly selected group of students in the halls.
   Ⓓ Joe & Mary's friends.

2. Which of the following represents who you should survey in a population?

   Ⓐ A random, representative group from the population
   Ⓑ Every individual in a population
   Ⓒ Only those in the population that agree with you
   Ⓓ Anyone, including those not in the population

3. What does increasing the sample size of a survey do for the overall results?

   Ⓐ Decreases bias in the results
   Ⓑ Increases the mean of the results
   Ⓒ Increases the reliability of the results
   Ⓓ Increasing sample size does not impact the results of a survey

4. Which of the following is not a valid reason for not surveying everyone in a population?

   Ⓐ It takes a far longer amount of time to survey everyone.
   Ⓑ Not everyone will be willing to participate in the survey.
   Ⓒ It is hard to determine the exact size of a population necessary to ensure everyone is surveyed.
   Ⓓ Surveying everyone produces unreliable results.

5. **Why is it important to know the sample size of a given survey?**

    Ⓐ It helps determine whether any bias exists.
    Ⓑ It helps determine how reliable the results are.
    Ⓒ It gives a good estimate for the size of the target population.
    Ⓓ It is not important to know the sample size.

6. **John and Maggie want to calculate the average height of students in their school. Which of the following groups of students would most likely produce the most amount of bias?**

    Ⓐ Every student in the 8th grade.
    Ⓑ Every student on the school basketball team.
    Ⓒ A randomly selected group of students in the halls.
    Ⓓ John & Maggie's friends.

7. **Which of the following question types will provide the most useful statistical results?**

    Ⓐ Open-ended questions where the person surveyed can answer in any way they want
    Ⓑ Multiple choice questions offering the person a representative number of choices
    Ⓒ True or false questions

8. **Which of the following does not represent a way of avoiding bias in survey results?**

    Ⓐ Use neutral words in the questions asked
    Ⓑ Ensure a random sample of the population
    Ⓒ Only survey individuals that will answer a certain way
    Ⓓ Tailor the conclusions based on survey results, not previous thoughts

9. **The following data set represents survey results on a scale of 1 to 10.**

    {8, 8, 9, 8, 6, 7, 7, 7, 8, 8, 6}

    Which of the following survey result would you be most surprised with if given by the next person surveyed?

    Ⓐ 6
    Ⓑ 5
    Ⓒ 8
    Ⓓ 7

10. **The following data set represents survey results on a scale of 1 to 10.**

{6, 6, 7, 6, 8, 7, 7, 7, 6, 6, 8}

Which of the following survey results would you be most surprised with if given by the next person surveyed?

Ⓐ 6
Ⓑ 10
Ⓒ 8
Ⓓ 7

11. **A company is conducting a survey on their performance with customer service. What method will best avoid receiving biased data?**

Ⓐ The company should make the survey anonymous.
Ⓑ The company should take a video survey.
Ⓒ The company should require the survey during the transaction.
Ⓓ The company should just request that everyone fill out a survey.

12. **A company is conducting a survey on their performance with customer service. What should the survey look like?**

Ⓐ The survey should contain a few simple multiple choice questions with an optional comment section.
Ⓑ The survey should contain simple free response questions with an optional comment section.
Ⓒ The survey should contain only free response questions.
Ⓓ The survey should contain some personal questions.

13. **Where should a survey about personal items purchased be conducted?**

Ⓐ The receipt should have a link to the survey website.
Ⓑ The survey should be near the register.
Ⓒ The survey should be near the exit doors.
Ⓓ The survey should be in the parking lot.

14. **How should online video game surveys be conducted?**

Ⓐ This survey should be put at the end of a level or section of the game.
Ⓑ This survey should be put at the beginning of the game.
Ⓒ Thus survey should be put on flyers and distributed at a gaming store.
Ⓓ This survey should be put on the website separate from the game.

15. **A large corporation is launching a new product, Hitz, which allows customers to purchase music and store it on the corporation's servers. They want to survey people who listen to a lot of music. Which of the following would give them the best sample?**

Ⓐ Conducting a survey inside of a music store

Ⓑ Conducting a survey at the parking lot of a music concert

Ⓒ Conducting an online survey on music social networking websites

Ⓓ Conducting a survey of students in a school marching band.

Describing Multiple Samples (7.SP.A.2)

1. **John comes up with the following methods for generating unbiased samples from shoppers at a mall.**

   I. Ask random strangers in the mall

   II. Always go to the mall at the same time of day

   III. Go to different places in the mall

   IV. Don't ask questions the same way to different people

   Which of these techniques represents the best way of generating an unbiased sample?

   - (A) I and II
   - (B) I and III
   - (C) I, II, and III
   - (D) All of these

2. **These two samples are about students' favorite subjects. What inference can you make concerning the students' favorite subjects?**

Student samples	Science	Math	English Language Arts	Total
#1	40	14	30	84
#2	43	17	33	93

   - (A) Students prefer Science over the other subjects.
   - (B) Students prefer Math over the other subjects.
   - (C) Students prefer English language arts over the other subjects.
   - (D) Students prefer History over the other subjects.

3. **These two samples are about students' favorite types of movies. What inference can you make concerning the students' favorite types of movies?**

Student samples	Comedy	Action	Drama	Total
#1	35	45	19	99
#2	38	48	22	108

   - (A) Students prefer action movies over the other types.
   - (B) Students prefer drama over the other types.
   - (C) Students prefer comedy over the other types.
   - (D) none

4. These two samples are about students' favorite fruits. What inference can you make concerning the students' favorite fruits?

Student samples	Blueberries	Bananas	Strawberries	Total
#1	33	18	44	95
#2	30	20	40	90

Ⓐ Students prefer strawberries over the other fruits.
Ⓑ Students prefer bananas over the other fruits.
Ⓒ Students prefer blueberries over the other fruits.
Ⓓ none

5. Jane and Matt conducted two surveys about students' favorite sports to play. What inference can you make concerning the students' favorite sports?

Student samples	Soccer	Basketball	Tennis	Total
#1	50	145	26	221
#2	56	150	20	226

Ⓐ Most students like basketball over soccer or tennis.
Ⓑ Most students like soccer over basketball or tennis.
Ⓒ Most students like tennis over soccer or basketball.
Ⓓ Most students like track.

6. Paul and Maggie conducted two surveys about students' favorite seasons. What inference can you make concerning the students' favorite seasons?

Student samples	Summer	Fall	Spring	Total
#1	100	128	244	472
#2	98	129	250	477

Ⓐ Most students prefer the Spring season over Summer and Fall.
Ⓑ Students prefer the Summer season.
Ⓒ Students prefer the Fall season.
Ⓓ Students prefer the Winter season.

7. **John and Mark conducted two surveys about students' favorite pizza toppings. What inference can you make concerning the students' favorite pizza toppings?**

Student samples	pineapple	pepperoni	extra cheese	Total
#1	145	237	118	500
#2	150	230	120	500

Ⓐ Most students like pepperoni.
Ⓑ Most students like extra cheese.
Ⓒ Most students like pineapple.
Ⓓ The students like pepperoni over pineapple or extra cheese.

8. **Jon and Minny conducted two surveys about students' favorite board games. What inference can you make concerning the students' favorite board games?**

Student samples	Game X	Game Y	Game Z	Total
#1	45	5	6	56
#2	50	6	2	58

Ⓐ Most students like Game Z.
Ⓑ Most students like **Game X** over **Game Y** or Game Z.
Ⓒ Most students like **Game Y**.
Ⓓ Most students don't like **Game X**.

9. **Tom and Bob conducted two surveys about their co-workers' favorite hobbies. What inference can you make concerning their co-workers' favorite hobbies?**

Co-woker sample	Play Golf	Play Video Games	Fishing	Total
#1	125	80	126	331
#2	118	83	130	331

Ⓐ Most of their coworkers spend a lot of money.
Ⓑ Most of their coworkers spend a lot of time inside.
Ⓒ Most of their coworkers spend a lot of time outside.
Ⓓ Most of their coworkers live in big houses.

10. **Bill and Jill conducted two surveys about students' favorite card games. What inference can you make concerning the students' favorite card games?**

Student samples	Hearts	Go fish	Spades	Total
#1	14	90	11	115
#2	10	88	14	122

Ⓐ Most students like speed.
Ⓑ Most students like hearts.
Ⓒ Most students like spades.
Ⓓ Most students like Go fish over hearts or spades.

11. **Moe and Lonnie conducted two surveys about students' favorite soda flavor. What inference can you make concerning the students' favorite soda flavor?**

Student samples	Stawberry	Orange	Rootbeer	Total
#1	10	11	88	109
#2	14	14	90	118

Ⓐ Most students like strawberry.
Ⓑ Most students like orange.
Ⓒ Most students like root beer over strawberry or orange.
Ⓓ Most students like chocolate.

12. **Billy and Larry conducted two surveys about students' favorite subject. What inference can you make concerning the students' favorite subject?**

Student samples	Math	English Language Arts	Science	Total
#1	111	111	111	333
#2	114	114	114	342

Ⓐ Most students like science.
Ⓑ Most students like math.
Ⓒ The same number of students like the three subjects equally.
Ⓓ Most students like English Language Arts.

13. Tom and Jerry conducted two surveys about students' favorite ice cream flavor. What inference can you make concerning the students' favorite ice cream flavor?

Student samples	Chocolate	Strawberry	Cookies-n-Cream	Total
#1	59	3	12	74
#2	61	4	15	80

Ⓐ Most students' like cookies-n-cream.
Ⓑ Most students like chocolate over strawberry or cookies-n-cream.
Ⓒ Most students' like strawberry.
Ⓓ Most students' like strawberry and cookies-n-cream.

14. Paul and Maggie conducted two surveys about students' favorite holiday. What inference can you make concerning the students' favorite holiday?

Student samples	Christmas	Thanksgiving	Easter	Total
#1	100	128	244	472
#2	98	129	250	477

Ⓐ Most students like Halloween.
Ⓑ Most students like Christmas.
Ⓒ Most students like Thanksgiving.
Ⓓ The students like Easter over Christmas or Thanksgiving.

15. Harper wants to conduct two surveys at his school, where 1,600 students attend. For the first survey, he wants to find out what types of cell phones the students in his school use. Of those students, he wants to survey them to determine their favorite apps. Which of the following is the best method for selecting a random sample?

Ⓐ Select 20 students from each first period class.
Ⓑ Select random students as they enter or exit school.
Ⓒ Select students based on how many text messages they make each month.
Ⓓ Select all of the students.

Mean, Median, and Mean Absolute Deviation (7.SP.B.3)

1. **Consider the following dot-plot for Height versus Weight.**

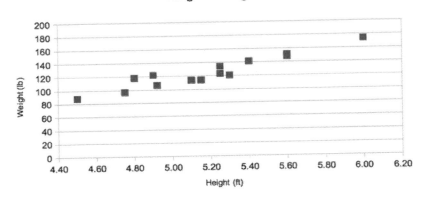

   What does this dot-plot indicate about the correlation between height and weight?

   - Ⓐ There is no correlation.
   - Ⓑ There is a strong negative correlation.
   - Ⓒ There is a strong positive correlation.
   - Ⓓ There is a weak positive correlation.

2. **The following chart represents the heights of boys on the basketball and soccer teams.**

Basketball	Soccer
5'4''	4'11''
5'2''	4'10''
5'3''	5'9''
5'5''	5'1''
5'5''	5'0''
5'1''	5'1''
5'9''	5'3''
5'3''	5'1''

   What inference can be made based on this information?

   - Ⓐ Soccer players have a higher average skill level than basketball players.
   - Ⓑ Soccer players have a lower average weight than basketball players.
   - Ⓒ Basketball players have a higher average height than soccer players.
   - Ⓓ No inference can be made.

3. **Use the table below to answer the question that follows:**

Month	Avg Temp.
January	24°F
February	36°F
March	55°F
April	65°F
May	72°F
June	78°F

What is difference between the mean temperature of the first four months of the year and the mean temperature of the next two months?

Ⓐ 15 degrees
Ⓑ 20 degrees
Ⓒ 25 degrees
Ⓓ 30 degrees

4. **Use the table below to answer the question:**

Month	Avg Temp.
January	24°F
February	36°F
March	55°F
April	65°F
May	72°F
June	78°F

If the temperature in January was 54°F instead of 24°F, by how much would the mean temperature for the six months increase?

Ⓐ 5°F
Ⓑ 10°F
Ⓒ 30°F
Ⓓ 35°F

5. **Use the table to answer the question below:**

Team	Wins
Mustangs	14
Spartans	17
North Stars	16
Hornets	9
Stallions	13
Renegades	9
Rangers	5

What is the mode of the wins for all the teams in the above table?

Ⓐ 14
Ⓑ 5
Ⓒ 13
Ⓓ 9

6. **Jack scored 7, 9, 2, 6, 15 and 15 points in 6 basketball games. Find the mean, median and mode scores for all the games.**

Ⓐ Mean = 7, Median = 9 and Mode = 2
Ⓑ Mean = 9, Median = 7 and Mode = 15
Ⓒ Mean = 9, Median = 8 and Mode = 15
Ⓓ Mean = 7, Median = 2 and Mode = 9

7. **Mean absolute deviation is a measure of...**

Ⓐ Central Tendency
Ⓑ Variability
Ⓒ Averages
Ⓓ Sample Size

8. Calculate the mean for the following set of data:

$$\left\{ \frac{1}{4}, \frac{3}{4}, \frac{5}{4}, \frac{7}{4}, \frac{3}{4}, \frac{11}{4} \right\}$$

Ⓐ $\dfrac{7}{4}$

Ⓑ $\dfrac{5}{2}$

Ⓒ $\dfrac{5}{4}$

Ⓓ $\dfrac{1}{6}$

9. Another word for mean is...

Ⓐ Average
Ⓑ Middle
Ⓒ Most
Ⓓ Count

10. What is the median for the following set of data?

$$\left\{ \frac{4}{5}, \frac{1}{3}, \frac{1}{3}, \frac{1}{5}, \frac{2}{3} \right\}$$

Ⓐ $\dfrac{1}{3}$

Ⓑ $\dfrac{2}{3}$

Ⓒ $\dfrac{1}{5}$

Ⓓ $\dfrac{1}{4}$

11. What is the mode of the following set of data?

{1, 1, 2, 3, 1, 4, 6, 2, 3}

Ⓐ 2
Ⓑ 3.75
Ⓒ 1
Ⓓ 6

12. John scored 6, 8, 1, 5, 11, 14 and 14 points in 7 lacrosse games. Find the mean, median and mode scores for all the games. (Round to the nearest tenth)

Ⓐ Mean = 8.4, Median = 8 and Mode = 14
Ⓑ Mean = 8.1, Median = 6, and Mode = 14
Ⓒ Mean = 9.3, Median = 8 and Mode = 15
Ⓓ Mean = 7.5, Median = 2 and Mode = 9

13. Calculate the mean for the following set of data:

{ 0.3, 1.2, 2.5, 4.3 }
Round your answer to the nearest tenth.

Ⓐ 2.1
Ⓑ 2.7
Ⓒ 8.3
Ⓓ 2.5

14. Calculate the median for the following set of data:

{ 0.3, 1.2, 2.5, 4.3 }
Round your answer to the nearest tenth.

Ⓐ 1.8
Ⓑ 2.1
Ⓒ 1.9
Ⓓ 1.2

15. What is the mean absolute deviation for the following set of data?

{2, 5, 7, 1, 2}

Ⓐ 3.4
Ⓑ 2
Ⓒ 2.08
Ⓓ 5

Mean, Median, and Mode (7.SP.B.4)

1. The following data set represents a score from 1-10 for a customer's experience at a local restaurant.

    { 1, 1, 2, 1, 3, 4, 7, 8, 1, 3, 4, 2, 1, 3, 7, 2 }

    If a score of 1 means the customer did not have a good experience, and a 10 means the customer had a fantastic experience, what can you infer by looking at the data?

    Ⓐ Overall, customers had a good experience.
    Ⓑ Overall, customers had a bad experience.
    Ⓒ Overall, customers had an "ok" experience.
    Ⓓ Nothing can be inferred from this data.

2. The manager of a local pizza place has asked you to make suggestions on how to improve his menu. The following bar graph represents the results of a survey asking customers what their favorite food at the restaurant was.

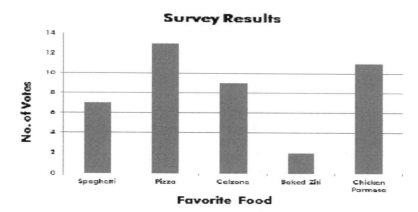

    Based on these survey results, which menu item would you suggest the manager remove from the menu?

    Ⓐ Spaghetti
    Ⓑ Pizza
    Ⓒ Calzone
    Ⓓ Baked Ziti

3. What are the measures of central tendency?

    Ⓐ Mean, Median, Mode
    Ⓑ Median, Mode, Mean Absolute Deviation
    Ⓒ Median, Mean Absolute Deviation, Sample Size
    Ⓓ Mean, Median, Range

4. **What is the mean absolute deviation for the following set of data?**

   {1, 2, 3, 4}

   Ⓐ 1
   Ⓑ 2.5
   Ⓒ 4
   Ⓓ 2

5. **What are the central tendencies of the following data set? (round to the nearest tenth)**

   {2, 2, 3, 4, 6, 8, 9, 10, 13, 13, 16, 17}

   Ⓐ Mean: 8.6, Median: 8.5, Mode: 2, 13
   Ⓑ Mean: 8.5, Median: 8.4, Mode: 2, 3
   Ⓒ Mean: 8.5, Median: 8.6, Mode: 2, 13
   Ⓓ Mean: 7.6, Median: 7.5, Mode: 10, 13

6. **What are the central tendencies of the following data set? (round to the nearest tenth)**

   {11, 11, 12, 13, 15, 17, 18, 20, 23, 23, 26, 27}

   Ⓐ Mean: 17.2, Median: 17.5, Mode: none
   Ⓑ Mean: 18.5, Median: 16, Mode: 11
   Ⓒ Mean: 17.5, Median: 18, Mode: 23
   Ⓓ Mean: 18, Median: 17.5, Mode: 11, 23

7. **What are the central tendencies of the following data set? (round to the nearest tenth)**

   {31, 31, 32, 33, 35, 37, 38, 41, 54, 54, 57, 58}

   Ⓐ Mean: 41.4, Median: 37.5, Mode: 31
   Ⓑ Mean: 41.5, Median: 36.5, Mode: none
   Ⓒ Mean: 41.6, Median: 38.5, Mode: 31, 54
   Ⓓ Mean: 41.8, Median: 37.5, Mode: 31, 54

8. **What are the central tendencies of the following data set? (round to the nearest tenth)**

   {41, 41, 42, 43, 45, 47, 48, 51, 64, 64, 64, 67}

   Ⓐ Mean: 51.4, Median: 47.5, Mode: 64
   Ⓑ Mean: 48.5, Median: 46, Mode: 64
   Ⓒ Mean: 47.5, Median: 48, Mode: 64
   Ⓓ Mean: 47, Median: 47.5, Mode: 64

9. **What are the central tendencies of the following data set? (round to the nearest tenth)**

{51, 51, 52, 53, 55, 57, 58, 61, 54, 54, 57, 54}

Ⓐ Mean: 58.5, Median: 56, Mode: 54
Ⓑ Mean: 54.8, Median: 54, Mode: 54
Ⓒ Mean: 57.5, Median: 58, Mode: 54
Ⓓ Mean: 57.6, Median: 57.5, Mode: 54

10. **What are the central tendencies of the following data set? (round to the nearest tenth)**

{61, 61, 52, 53, 65, 67, 58, 61, 64, 64, 57, 54}

Ⓐ Mean: 59.8, Median: 61, Mode: 61
Ⓑ Mean: 58.5, Median: 66, Mode: 64
Ⓒ Mean: 57.5, Median: 68, Mode: 64
Ⓓ Mean: 57.9, Median: 67.5, Mode: 64

11. **What are the central tendencies of the following data set? (round to the nearest tenth)**

{71, 71, 62, 63, 75, 77, 68, 71, 74, 74, 67, 74}

Ⓐ Mean: 70.6, Median: 78, Mode: 74
Ⓑ Mean: 68.5, Median: 76, Mode: 74
Ⓒ Mean: 70.6, Median: 71, Mode: 74
Ⓓ Mean: 77, Median: 77.5, Mode: 74

12. **What are the central tendencies of the following data set? (round to the nearest tenth)**

{61, 71, 52, 63, 65, 77, 58, 71, 64, 74, 57, 74}

Ⓐ Mean: 65.6, Median: 64.5, Mode: 71, 74
Ⓑ Mean: 68.5, Median: 66, Mode: none
Ⓒ Mean: 67.5, Median: 68, Mode: 71
Ⓓ Mean: 67, Median: 67.5, Mode: 2

13. **What are the central tendencies of the following data set? (round to the nearest tenth)**

{21, 21, 22, 23, 35, 37, 38, 41, 44, 44, 47, 44}

Ⓐ Mean: 38.5, Median: 37.5, Mode: 44
Ⓑ Mean: 34.8, Median: 37.5, Mode: 44
Ⓒ Mean: 37.5, Median: 38, Mode: none
Ⓓ Mean: 37, Median: 37.5, Mode: 54

14. **Kelli's Ice Cream Shop must have a mean of 110 visitors per day in order to make a profit. The table below shows the number of visitors during one week.**

   Based on the table, how many visitors will need to go to Kelli's Ice Cream shop on Saturday in order for the company to make a profit for the week?

   - Ⓐ 108
   - Ⓑ 110
   - Ⓒ 122
   - Ⓓ 222

15. **Henry made a list of his math scores for one grading period: 98, 87, 93, 89, 90, 85, 88. If Henry adds a 90 to the list to represent his final test for the grading period, which statement is true?**

   - Ⓐ The mode would decrease.
   - Ⓑ The mean would increase.
   - Ⓒ The median would increase.
   - Ⓓ The mean would decrease.

Understanding Probability (7.SP.C.5)

1. Mary has 3 red marbles and 7 yellow marbles in a bag. If she were to remove 2 red and 1 yellow marbles and set them aside, what is the probability of her pulling a yellow marble as her next marble?

    (A) $\dfrac{1}{6}$

    (B) $\dfrac{1}{7}$

    (C) $\dfrac{7}{10}$

    (D) $\dfrac{6}{7}$

2. John has a deck of cards (52 cards). If John removes a number 2 card from the deck, what is the probability that he will pick a number 2 card at random?

    (A) 3 out of 51
    (B) 4 out of 51
    (C) 26 out of 51
    (D) 30 out of 51

3. Maggie has a bag of coins (8 nickels, 6 quarters, 12 dimes, 20 pennies). If she picks a coin at random, what is the probability that she will pick a quarter?

    (A) 2 out of 15
    (B) 3 out of 23
    (C) 3 out of 50
    (D) 5 out of 46

4. **Mark has a box of bills (12 ones, 8 tens, 21 twenties, 30 fifties). If he picks a bill at random, what is the probability that he will pick a ten?**

    Ⓐ 8 out of 71
    Ⓑ 8 out of 100
    Ⓒ 10 out of 71
    Ⓓ 7 out of 71

5. **Moe has a bowl of nuts (14 pecans, 8 walnuts, 28 almonds, 33 peanuts). If he picks a nut at random, what is the probability that he will pick a peanut?**

    Ⓐ 33 out of 70
    Ⓑ 33 out of 80
    Ⓒ 33 out of 100
    Ⓓ 33 out of 83

6. **Xavier has a bowl of nuts (14 pecans, 8 walnuts, 28 almonds, 33 peanuts). If he picks out all the pecans, what is the probability that he will pick a walnut at random?**

    Ⓐ 8 out of 83
    Ⓑ 8 out of 69
    Ⓒ 8 out of 100
    Ⓓ 8 out of 70

7. **Tim has a box of chocolates with the following flavors: 24 cherry, 26 caramel, 20 fudge, and 16 candy. If he picks out two of each type of chocolate, what is the probability that he will pick a cherry chocolate at random?**

    Ⓐ 11 out of 35
    Ⓑ 4 out of 13
    Ⓒ 11 out of 50
    Ⓓ 11 out of 39

8. **Clarissa has a box of chocolates with the following flavors: 24 cherry, 26 caramel, 20 fudge, and 16 taffy. If she picks out all the taffy, what is the probability that she will pick a fudge chocolate at random?**

    Ⓐ 2 out of 7
    Ⓑ 10 out of 43
    Ⓒ 1 out of 5
    Ⓓ 8 out of 35

9. Karen has a box of chocolates with the following flavors: 24 cherry, 26 caramel, 20 fudge, and 16 taffy. If she removes half of the cherry and fudge chocolates from the box, what is the probability that she will pick a taffy chocolate at random?

Ⓐ 1 out of 4
Ⓑ 8 out of 43
Ⓒ 4 out of 25
Ⓓ 16 out of 43

10. The table below shows the types of fruits Jessie's mom purchased from the grocery store.

Type of Fruit	Number of Fruits
Apple	6
Orange	4
Pear	2
Peaches	3

If Jessie grabs one of the fruits without looking, what is the probability that he will NOT pick a peach?

Ⓐ $\dfrac{1}{5}$

Ⓑ $\dfrac{3}{15}$

Ⓒ $\dfrac{4}{5}$

Ⓓ $\dfrac{2}{3}$

Predicting Using Probability (7.SP.C.6)

1. Which of the following represents the sample space for flipping two coins?

    Ⓐ {HH, TT}
    Ⓑ {H, T}
    Ⓒ {HH, HT, TH, TT}
    Ⓓ {HH, HT, TT}

2. Which of the following experiments would best test the statement, "The probability of a coin landing on heads is 1/2."?

    Ⓐ Toss a coin 1,000 times, and record the results.
    Ⓑ Toss a coin twice to see if it lands on heads one out of those two times.
    Ⓒ Toss a coin until it lands on heads and record the number of tries it took.
    Ⓓ Toss a coin twice, if it doesn't land on heads exactly once, the theoretical probability is false.

3. Which of the following results is most likely from tossing a six-sided die?

    Ⓐ Rolling an odd number
    Ⓑ Rolling an even number
    Ⓒ Rolling a number from 1 to 3
    Ⓓ All of the above are equally likely.

4. Sandy flipped a coin 40 times. Her results are 75% heads and 25% tails. What is the difference between the actual results and the expected results?

    Ⓐ 20%
    Ⓑ 50%
    Ⓒ 25%
    Ⓓ 10%

5. Maggie rolled a pair of four sided dice 10 times. The results are 30% side 1, 20% side 2, 20% side 3, 30% side 4. What is the difference between the results and the expected results for all four sides?

Ⓐ 25% side 1,
  25% side 2,
  25% side 3,
  25% side 4

Ⓑ 5% side 1,
  5% side 2,
  5% side 3,
  5% side 4

Ⓒ 5% side 1,
  25% side 2,
  25% side 3,
  5% side 4

Ⓓ 4% side 1,
  4% side 2,
  4% side 3,
  4% side 4

6. Maggie rolls two pairs of four sided dice 10 times. The results were 30% side 1, 20% side 2, 20% side 3, 30% side 4. What were the actual results and expected results?

Ⓐ Results: 12 side 1, 8 side 2, 8 side 3, 12 side 4 ...Expected Results: 10 side 1, 10 side 2, 10 side 3, 10 side 4

Ⓑ Results: 6 side 1, 4 side 2, 4 side 3, 6 side 4 ...Expected Results: 5 side 1, 5 side 2, 5 side 3, 5 side 4

Ⓒ Results: 6 side 1, 4 side 2, 4 side 3, 6 side 4 ...Expected Results: 10 side 1, 10 side 2, 10 side 3, 10 side 4

Ⓓ Results: 10 side 1, 10 side 2, 10 side 3, 10 side 4 ...Expected Results: 12 side 1, 8 side 2, 8 side 3, 12 side 4

7. Juliana was expected to make 80% of her first serves in the tennis match. She made half of 60 first serves in the match. What were the results and expected results?

Ⓐ Results: 30 first serves. Expected Results: 48 first serves
Ⓑ Results: 20 first serves. Expected Results: 48 first serves
Ⓒ Results: 25 first serves. Expected Results: 48 first serves
Ⓓ Results: 35 first serves. Expected Results: 48 first serves

8. **Which of the following represents the sample space for rolling a pair of four-sided dice?**

    Ⓐ (1,1) (1,2) (1,3) (1,4) (2,2) (2,3) (2,4) (3,3) (3,4) (4,4)
    Ⓑ (1,2) (1,3) (1,4) (2,3) (2,4) (3,4)
    Ⓒ (1,2) (1,3) (1,4) (2,3) (2,4)(3,4) (4,4)
    Ⓓ (1,2) (1,3) (1,4) (2,3) (2,4)

9. **Lea is playing a carnival game and has a chance to win a large stuffed teddy bear. In order to win, she has to guess the color of cube she will pick out of a box. The box is shown below.**

    "Pick the Cube" Carnival Game

    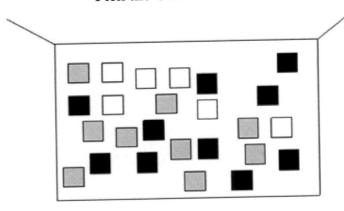

    Based on the colors and number of blocks, what is Lea's chance of winning the game if she says that she will pick a black cube out of the box?

    Ⓐ 60%
    Ⓑ 10%
    Ⓒ 90%
    Ⓓ 40%

10. **Ms. Green is passing out snacks during her tutoring session. She has 7 bags of chips, 13 candy bars, 16 fruit snacks, and 10 lollipops. If one of Ms. Green's students randomly selects a snack, what is the probability that he will select a fruit snack? Round your answer to the nearest tenth of a percent.**

    Ⓐ 34.8%
    Ⓑ 65.2%
    Ⓒ 16.0%
    Ⓓ 84.0%

Using Probability Models (7.SP.C.7.A)

1. Sara rolls two dice, one black and one yellow. What is the probability that she will roll a 3 on the black die and a 5 on the yellow die?

   Ⓐ $\dfrac{1}{6}$

   Ⓑ $\dfrac{1}{12}$

   Ⓒ $\dfrac{2}{15}$

   Ⓓ $\dfrac{1}{36}$

2. Which of the following represents the probability of an event most likely to occur?

   Ⓐ 0.25
   Ⓑ 0.91
   Ⓒ 0.58
   Ⓓ 0.15

3. Which of the following is not a valid probability?

   Ⓐ 0.25

   Ⓑ $\dfrac{1}{5}$

   Ⓒ 1

   Ⓓ $\dfrac{5}{4}$

4. Tom tosses a coin 12 times. The coin lands on heads only twice. If Tom tosses the coin one more time, what is the probability that the coin will land on heads?

   Ⓐ 40%
   Ⓑ 50%
   Ⓒ 60%
   Ⓓ 15%

5. Joe has 5 nickels, 5 dimes, 5 quarters, and 5 pennies in his pocket. Six times, he randomly picked a coin from his pocket and put it back. Joe picked a dime every time. If he randomly picks a coin from his pocket again, what is the probability the coin will be a dime?

Ⓐ 33%
Ⓑ 100%
Ⓒ 24%
Ⓓ 25%

6. Jim rolls a pair of six-sided dice five times. He rolls a pair of two's five times in a row. If he rolls the dice one more time, what is the probability he will roll a pair of fours?

Ⓐ 1 out of 21
Ⓑ 1 out of 36
Ⓒ 1 out of 24
Ⓓ 1 out of 6

7. Bob rolls a six-sided die and flips a coin five times. He rolls a three and flips the coin to tails five times in a row. If he rolls the die and flips the coin one more time, what is the probability he will roll a three and flip the coin on tails?

Ⓐ 1 out of 6
Ⓑ 1 out of 10
Ⓒ 1 out of 12
Ⓓ 1 out of 7

8. Mia rolls a pair of six-sided dice and flips two coins ten times. She rolls a pair of threes and flips the coins to tails ten times in a row. If she rolls the dice and flips the coins one more time, what is the probability she will roll all fives and flip the coins on heads?

Ⓐ 1 out of 60
Ⓑ 1 out of 144
Ⓒ 1 out of 128
Ⓓ 1 out of 64

9. Sophia wants to select a pair of shorts from Too Sweet Clothing Store. The store has 2 different colors of shorts (black (B) and green (G)) available in sizes of small (S), medium (M), and large (L). If Sophia grabs a pair of shorts without looking, which sample space shows the different types of shorts she could select?

Ⓐ {BS, BM, BL, GS, GM, GL}
Ⓑ {BB, BG, SS, SM, SL}
Ⓒ {BG, SM, SL}
Ⓓ {BS, BL, GS, GL}

10. Eric's Pizza Shop is offering a sale on any large two-topping pizza for $8.00. Customers have a choice between thin crust or pan pizza, one meat topping (pepperoni, Italian sausage, or ham), and one vegetable topping (green peppers, onions, or mushrooms). Which sample space shows the number of possible combinations customers can have if they select a thin crust pizza?

Ⓐ {TPG, TIO, THM}

Ⓑ {TPG, TPO, TPM, TIG, TIO, TIM, THG, THO, THM}

Ⓒ {TPG, TPO, TIG, TIO, THG, THO}

Ⓓ {TPG, TPO, TPM, TIG, TIO, TIM, THG, THO, THM, PPG, PPO, PPM, PIG, PIO, PIM, PHG, PHO, PHM}

1. Felix flipped a coin 8 times and got the following results: H, H, T, H, H, T, T, H. If these results were typical for that coin, what are the odds of flipping a heads with that coin?

   Ⓐ 3 out of 5
   Ⓑ 5 out of 8
   Ⓒ 3 out of 8
   Ⓓ 1 out of 2

2. Bridgette rolled a six-sided die 100 times to test the frequency of each number's appearing. According to these statistics, how many times should a 2 be rolled out of 50 rolls?

Number	Frequency
1	18%
2	20%
3	16%
4	11%
5	18%
6	17%

   Ⓐ 10 times
   Ⓑ 20 times
   Ⓒ 12 times
   Ⓓ 15 times

3. Randomly choosing a number out of a hat 50 times resulted in choosing an odd number a total of four more times than the number of times an even number was chosen. How many times was an even number chosen from the hat?

   Ⓐ 27 times
   Ⓑ 21 times
   Ⓒ 29 times
   Ⓓ 23 times

4. 8 out of the last 12 customers at Paul's Pizza ordered pepperoni pizza. According to this data, what is the probability that the next customer will NOT order pepperoni pizza?

   Ⓐ 1 out of 3
   Ⓑ 4 out of 5
   Ⓒ 1 out of 2
   Ⓓ 2 out of 5

5. Susan is selling cookies for a fundraiser. Out of the last 20 people she asked, 10 people bought 1 box of cookies, 5 bought more than 1 box, and 5 bought none. Based on this data, what is the probability that the next person she asks will buy at least 1 box of cookies?

Ⓐ 1 out of 3

Ⓑ 2 out of 5

Ⓒ 4 out of 5

Ⓓ 3 out of 4

6. **William has passed 9 out of his last 10 tests in Spanish class. Based on his past history, what is the probability that he will NOT pass the next test?**

Ⓐ 10%

Ⓑ 25%

Ⓒ 15%

Ⓓ 8%

7. **Travis has scored goals in 7 of his last 9 soccer games. At this rate, what is the probability that he will score in his next game? Round the nearest percent.**

Ⓐ 70%

Ⓑ 17%

Ⓒ 78%

Ⓓ 53%

8. **Gabe's free throw percentage for the season has been 80%. Based on this, if he has 5 free throws in the next game, how many is he likely to miss?**

Ⓐ 0

Ⓑ 1

Ⓒ 2

Ⓓ 3

9. **York and his partner have won the doubles tennis tournament three out of the last four years. According to this record, what is the probability they will win it again this year?**

Ⓐ 3 out of 4

Ⓑ 1 out of 3

Ⓒ 1 out of 2

Ⓓ 4 out of 5

10. **Robbie runs track. His finishes for his last 6 events were: 1st, 3rd, 2nd, 5th, 4th, 2nd. Based on these results, what is the probability he will finish in the top 3 of his next event?**

Ⓐ 4 out of 5

Ⓑ 2 out of 3

Ⓒ 1 out of 2

Ⓓ 3 out of 4

Find the Probability of a Compound Event (7.SP.C.8.A)

1. **The following tree diagram represents Jane's possible outfits:**

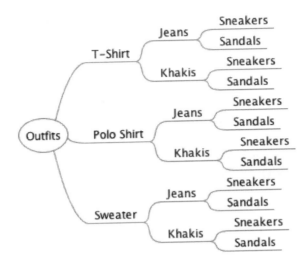

  How many different outfits can Jane make based on this diagram?

  Ⓐ 2
  Ⓑ 12
  Ⓒ 16
  Ⓓ 4

2. **The following tree diagram represents Jane's possible outfits:**

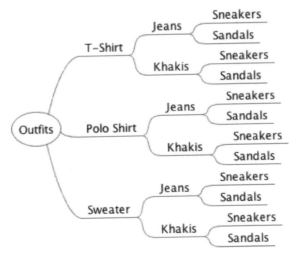

If Jane randomly selects an outfit, what is the probability she will be wearing Jeans AND Sneakers?

Ⓐ $\dfrac{1}{12}$

Ⓑ $\dfrac{1}{3}$

Ⓒ $\dfrac{1}{4}$

Ⓓ $\dfrac{1}{2}$

3. **Paul, Jack, Tom, Fred, and Sam are competing in the long jump. If they win the top five spots, how many ways could they be arranged in the top five spots?**

Ⓐ 15 ways
Ⓑ 60 ways
Ⓒ 120 ways
Ⓓ 3,125 ways

4. **Mona is about to roll a pair of four-sided dice. What is the probability she will roll a one and a two or a one and a three?**

Ⓐ 1 out of 8
Ⓑ 1 out of 16
Ⓒ 1 out of 12
Ⓓ 1 out of 4

5. **Tim rolls a pair of six-sided dice. What is the probability he will roll doubles?**

Ⓐ 6 out of 21
Ⓑ 1 out of 6
Ⓒ 1 out of 4
Ⓓ 2 out of 21

6. The triple jump competition is close. Joe, Damon, Sam, and Chris have a shot at first place. If two of the four tie for first place and the other two tie for second place, how many ways could they be arranged in the top two spots?

    Ⓐ 6 ways
    Ⓑ 2 ways
    Ⓒ 3 ways
    Ⓓ 8 ways

7. Sam is about to flip three coins. What is the probability he will flip all of the coins to heads?

    Ⓐ 1 out of 4
    Ⓑ 1 out of 6
    Ⓒ 1 out of 8
    Ⓓ 1 out of 2

8. Jona rolls one six-sided die and one four-sided die. What is the probability she will not roll a 2 or a 3 on either die?

    Ⓐ 2 out of 3
    Ⓑ 1 out of 3
    Ⓒ 1 out of 4
    Ⓓ 3 out of 4

9. Elsie rolled three four-sided dice. What is the probability she will roll one even and two odds?

    Ⓐ 1 out of 3
    Ⓑ 3 out of 8
    Ⓒ 1 out of 4
    Ⓓ 5 out of 8

10. What are the central tendencies of the following data set? (round to the nearest tenth)

    {21, 21, 22, 23, 25, 27, 28, 31, 34, 34, 34, 37}

    Ⓐ Mean: 28.1, Median: 27.5, Mode: 34
    Ⓑ Mean: 28.5, Median: 26, Mode: none
    Ⓒ Mean: 27.5, Median: 28, Mode: none
    Ⓓ Mean: 27, Median: 27.5, Mode: none

Represent Sample Spaces (7.SP.C.8.B)

1. **If Robbie flips a quarter twice, what is the sample space for the possible outcomes?**

   Ⓐ HT, HH, TT, TH
   Ⓑ HT, TH
   Ⓒ HT, TT, TH
   Ⓓ HH, TT

2. **If Bret rolls a six-sided die twice, which table shows the sample space for possible outcomes?**

Ⓐ

	1	2	3	4	5	6
1	1, 1	1, 2	1, 3	1, 4	1, 5	1, 6
2	2, 1	2, 2	2, 3	2, 4	2, 5	2, 6
3	3, 1	3, 2	3, 3	3, 4	3, 5	3, 6
4	4, 1	4, 2	4, 3	4, 4	4, 5	4, 6
5	5, 1	5, 2	5, 3	5, 4	5, 5	5, 6
6	6, 1	6, 2	6, 3	6, 4	6, 5	6, 6

Ⓑ

	1	2	3	4	5	6
1	1, 1					
2		2, 2				
3			3, 3			
4				4, 4		
5					5, 5	
6						6, 6

	1	2	3	4	5	6
1		1, 2	1, 3	1, 4	1, 5	1, 6
2	2, 1		2, 3	2, 4	2, 5	2, 6
3	3, 1	3, 2		3, 4	3, 5	3, 6
4	4, 1	4, 2	4, 3		4, 5	4, 6
5	5, 1	5, 2	5, 3	5, 4		5, 6
6	6, 1	6, 2	6, 3	6, 4	6, 5	

Ⓒ

	1	2	3	4
1	1, 1	1, 2	1, 3	1, 4
2	2, 1	2, 2	2, 3	2, 4
3	3, 1	3, 2	3, 3	3, 4
4	4, 1	4, 2	4, 3	4, 4
5	5, 1	5, 2	5, 3	5, 4
6	6, 1	6, 2	6, 3	6, 4

Ⓓ

3. **There are three colors of stones in a bag: red, green, and blue. Two stones are drawn out at random. What are the possible outcomes in which exactly one blue stone might be drawn?**

Ⓐ BR, GB BG

Ⓑ BG, BR, BB

Ⓒ RB, GB, BR, BG

Ⓓ BG, BR

4. A box contains both red checkers and black checkers. Four checkers are drawn out. How many different possible outcomes would result in exactly three checkers being red?

  Ⓐ 8
  Ⓑ 2
  Ⓒ 6
  Ⓓ 4

5. A number between 1 and 10 is chosen twice. How many different ways might the same number be chosen both times?

  Ⓐ 5
  Ⓑ 10
  Ⓒ 0
  Ⓓ 3

6. A number between 1 and 10 is chosen twice. The sample space for the results includes how many outcomes that are even numbers followed by prime numbers?

  Ⓐ 20
  Ⓑ 9
  Ⓒ 25
  Ⓓ 12

7. Two six-side dice are rolled. How many different outcomes for the two dice would result in a total of 7 being rolled?

  Ⓐ 10
  Ⓑ 8
  Ⓒ 6
  Ⓓ 11

8. A stack of 20 cards are numbered 1 through 20. If cards are drawn from the stack until the 3 card is drawn, how many different outcomes would result in the 3 being drawn on the third draw?

  Ⓐ 37
  Ⓑ 112
  Ⓒ 20
  Ⓓ 342

9. A target has 4 rings on it. How many different ways might two arrows be distributed on the target?

    Ⓐ 4
    Ⓑ 16
    Ⓒ 8
    Ⓓ 7

10. If five different players have to be placed in five different positions on the team, how many different ways might this be done?

    Ⓐ 120
    Ⓑ 15
    Ⓒ 40
    Ⓓ 75

Simulate Compound Events to Estimate Probability (7.SP.C.8.C)

1. If 20% of applicants for a job are female, what is the probability that the first two applicants will be male?

   Ⓐ 64%
   Ⓑ 80%
   Ⓒ 60%
   Ⓓ 52%

2. If you want to simulate a random selection from a large population that is 40% adult and 60% children, how can you use slips of paper to do so?

   Ⓐ Make 5 slips of paper, 2 for adults and 3 for children. Randomly select slips of paper from the 5 to represent the choice of someone from the population.
   Ⓑ Make 2 slips of paper, 1 for adults and 1 for children. Randomly select slips of paper from the 2 to represent the choice of someone from the population.
   Ⓒ Make 100 slips of paper, 50 for adults and 50 for children. Randomly select slips of paper from the 100 to represent the choice of someone from the population.
   Ⓓ Make 3 slips of paper, 1 for adults and 2 for children. Randomly select slips of paper from the 3 to represent the choice of someone from the population.

3. A sandwich shop has 6 breads and 5 meats available for sandwiches. What is the probability that two people in a row will chose the same combination of bread and meat?

   Ⓐ 1 out of 11
   Ⓑ 1 out of 2
   Ⓒ 1 out of 30
   Ⓓ 1 out of 20

4. A catalogue has sports uniforms for sale. There are 6 designs of shorts that can be combined with 4 designs of shirts. What is the probability that two teams choose different shorts and different shirts?

   Ⓐ 1 out of 2
   Ⓑ 5 out of 8
   Ⓒ 1 out of 4
   Ⓓ 2 out of 7

5. Sally has to choose a pair of pants and a pair of shoes to wear to her club meeting. She can't remember what she wore last time. She has 3 pairs of pants and 5 pairs of shoes to choose from. What is the probability that she will wear the same combination that she wore last time?

  Ⓐ 1 out of 15
  Ⓑ 1 out of 8
  Ⓒ 2 out of 11
  Ⓓ 1 out of 125

6. Suppose that each of the next 5 days there is a 50% chance of rain. You want to know the likelihood of it not raining at all in those 5 days. How can you test that probability with a coin?

  Ⓐ Flip the coin. If it is heads, there will be no rain, and if it is tails, there will be rain. Flip it at least 10 times and see how many times no rain is the result.
  Ⓑ Flip a coin until you get tails, which will represent rain. If you get rain in fewer than 5 coin flips, it will rain in the next five days.
  Ⓒ Flip a coin five times in a row. Repeat this numerous times. Let heads represent no rain and tales represent rain. See how often the five coin flips result in no rain.
  Ⓓ Flip a coin five times. Take the number of times that tails is flipped, and divide it by five. That will tell you the probability of rain in the next five days.

7. Your lawnmower starts well four times out of five. If you have to mow the lawn three more times this season, what is the probability that it will start well all three times?

  Ⓐ About 51%
  Ⓑ About 43%
  Ⓒ About 87%
  Ⓓ About 22%

8. Evan works with his friend Luke. His friend is a good worker, but he has a tendency to be late to work too often. He only shows up on time about 50% of the time. He has already been late once in the first two days of the work week. What is the probability that he will be on time for the remaining three days of the workweek?

  Ⓐ 1 in 5
  Ⓑ 1 in 2
  Ⓒ 1 in 8
  Ⓓ 1 in 6

9. A group of three friends was curious about which day of the week each of them was born on. They decided to research it to find out. What is the probability that all three of them were born on the same day of the week?

Ⓐ 1 out of 7
Ⓑ 1 out of 49
Ⓒ 1 out of 343
Ⓓ 1 out of 21

10. Two friends both happened to buy new trucks from the same manufacturer in the same week. The manufacturer offers 4 models of trucks in 6 different colors. What is the probability that the two friends happened to buy the same model and color of truck?

Ⓐ 1 out of 10
Ⓑ 1 out of 14
Ⓒ 1 out of 15
Ⓓ 1 out of 24

End of Statistics and Probability

Statistics and Probability

Answer Key
&
Detailed Explanations

Sampling a Population (7.SP.A.1)

Question No.	Answer	Detailed Explanation
1	C	A group of randomly selected students in the hallways would produce the least amount of bias because it is unlikely for assumptions to be made or factors that influence the data to be present. For example, students in the 8th grade may be taller than other students in other grades, skewing the data toward a higher average. A similar assumption can be made about students on the basketball team. Joe and Mary would already have an idea of the height of their friends.
2	A	A random, representative group represents the people to survey in a population because each person in the population has an equal chance of being included and there is less of a chance of bias altering the results of the survey.
3	C	As the size of the sample (people surveyed) increases, the results become more accurate. Therefore, increasing the sample size increases the reliability of the results.
4	D	Surveying everyone would produce reliable results because there would specific data from each person; however, due to the length of time needed, among other reasons, surveying everyone may not be possible (depending on the reasons for the survey).
5	B	A sample size gets information from a small group from the population. Since the sample size is supposed to represent the population at large, it's important to know the parameters of the sample size in order to determine how reliable the results are for a given survey.
6	B	An assumption can be made that students on the basketball team are likely to be taller than other students at the school, which can influence the results (make it appear that the average height of the general school population is higher than it actually is). Therefore, calculating the average height of the basketball team would most likely produce the most amount of bias.

Question No.	Answer	Detailed Explanation
7	B	Multiple choice questions are useful for statistical results because they can represents different amounts of data that can be separated into categories and sub-categories based on the type of survey and the intended outcome.
8	C	Bias in collecting data can be caused by using questions where there are likely to be assumptions made or factors that influence the data. A survey that asks questions that cause individuals to answer in a certain way makes assumptions and influences the outcome of the data. Therefore, using that type of survey does not avoid bias.
9	B	When analyzing the data, out of the 10 responses, none of the responses have been less than 6. Therefore, a surprising response from the next person would be a 5 since it is outside the range.
10	B	When analyzing the data, out of the 10 responses, all are within the range of 6 to 8. Therefore, a surprising response from the next person would be a 10 since it is outside of the range.
11	A	Bias in collecting data can be caused by using questions where there are likely to be assumptions made or factors that influence the data. An anonymous survey would encourage individuals to answer questions openly without worry about the implications of their answers. Therefore, an anonymous survey is the best way to avoid receiving biased data.
12	A	Multiple choice questions are useful for statistical results because they can represent different amounts of data that can be separated into categories and sub-categories based on the type of survey and the intended outcome. The optional comments section gives an individual the opportunity to make open observations (without bias). A survey with too many free response questions would be hard to analyze and personal questions invite bias.
13	A	For collecting personal data, an anonymous survey would encourage individuals to answer questions openly without worry about the implications of their answers. It is also a way to avoid receiving bias data. Therefore, a receipt should have a link to an online survey where individuals can enter information about personal data.

Question No.	Answer	Detailed Explanation
14	A	In order to get a reliable, random sample based on the population (players who use the game), the surveyors should set up their survey at the end of a level or section of the game. Players are more likely to answer the survey before going to the next level. People are ready to play at the beginning of a game and ready to turn the game off near the end, so there is a good chance they won't complete the survey. A separate survey is likely to be ignored.
15	C	When choosing a sample to survey for data, the sample should be representative of the target population. Here, the target population is people most likely to purchase music online. These people are most likely to communicate with others about music online. Therefore, an online survey on music social networking websites would give them the best sample.

Describing Multiple Samples (7.SP.A.2)

1	B	A group of randomly selected strangers in different places of a mall would produce the least amount of bias because there is unlikely to be assumptions made or factors that influence the data.
2	A	In the survey, 83 students selected science as their favorite subject while the other subjects had a combined number of 94 students. Therefore, an inference can be made that students prefer science over the other subjects.
3	A	In the sample, a combined 93 students chose action movies, 41 students chose drama, and 73 chose comedy. Therefore, an inference can be made that students prefer action movies to other types of movies.
4	A	In the sample, a combined 84 students chose strawberries, 38 students chose bananas and 63 chose blueberries. Therefore, an inference can be made that students prefer strawberries to other types of fruit.
5	A	In the sample, a combined 295 students chose basketball, 106 students chose soccer, and 46 students chose playing tennis. Since more students chose basketball, an inference can be made that most students like basketball over soccer or tennis.

Question No.	Answer	Detailed Explanation
6	A	In the sample, a combined 198 students chose summer, 257 students chose fall, and 494 students chose spring. Since more students chose spring, an inference can be that most students prefer the Spring season over Summer or Fall.
7	D	In the sample, a combined 295 students chose pineapple, 238 students chose extra cheese, and 467 students chose pepperoni. Since more students chose pepperoni, an inference can be that most students prefer the pepperoni over pineapple or extra cheese.
8	B	In the sample, a combined 95 students chose Game X, 11 students chose Game Y, and 8 students chose Game Z. Since more students chose Game X, an inference can be that most students like Game X over Game Y or Game Z.
9	C	In the sample, a combined 243 co-workers play golf, 163 co-workers chose video games, and 256 co-workers chose fishing. Since golf and fishing are outdoor activities, an inference can be that most of Tom and Bob's co-workers spend a lot of time outside.
10	D	In the sample, a combined 178 students chose Go fish, 24 students chose hearts, and 25 students chose spades. Since more students chose Go fish, an inference can be that most students like Go fish over hearts or spades.
11	C	In the sample, a combined 24 students chose strawberry, 178 students chose rootbeer, and 25 students chose orange. Since more students chose root beer, an inference can be that most students prefer root beer over strawberry or orange.
12	C	In the sample, a combined 225 students chose math, 225 students chose English and 225 students chose science. Since the students' answers reflect the same numbers for each subject, an inference can be made that the same number of students like the three subjects equally.
13	B	In the sample, a combined 120 students chose chocolate, 7 students chose strawberry and 27 students chose cookies-n-cream. Since more students chose chocolate, an inference can be made that most students like chocolate over strawberry or cookies-n-cream.

Question No.	Answer	Detailed Explanation
14	D	In the sample, a combined 198 students chose Christmas, 257 students chose Thanksgiving and 494 students chose Easter. Since more students chose Easter, an inference can be made that most students like Easter over Christmas or Thanksgiving.
15	B	A random, representative group represents the people to survey in a population because each person in the population has an equal chance of being included and there is less of a chance of bias altering the results of the survey. Therefore, Harper should select students as they enter and exit the school.

Mean, Median, and Mean Absolute Deviation (7.SP.A.2)

1	C	There is a strong positive correlation between the height and the weight because there is an upward trend in the weight as a person gets taller.
2	C	When analyzing the heights of the basketball players and soccer players, the average height of the basketball players is 5'4, and the average height of the soccer players is 5 1 1/2. Therefore, the average height of the basketball players is higher than the average height of the soccer players.
3	D	Remember: the mean represents the average of the values, which is calculated by adding all the values together then dividing by the number of values you added. In this case, (24 + 36 + 55 + 65) / 4 = 45, and (72 + 78) / 2 = 75. The difference between these two values is 30, as indicated.
4	A	Remember: A mean is the average of all the data presented. In this case, you have to average the temperatures, then replace the January temperature with 54 and recalculate the average, then take the difference.
5	D	Remember: the mode of a set of data is the number that occurs most frequently. For this set of data, that number is 9.
6	C	Remember: Mean is the average, median is the middle-number when data is arranged numerically, and mode is the number that appears most often.
7	B	Mean absolute deviation is a measure of the distance between each data value and the mean, so it measures variability.

Question No.	Answer	Detailed Explanation
8	C	To find the mean, add the values, and divide that value by the amount of numbers in the data set. (1) The sum is 30/4 (2) 30/4 ÷ 6/1 = (3) 30/4 × 1/6 = 30/24 (the GCF of 30 and 24 is 6) (4) 5/4
9	A	Mean is the average of a set of data; therefore, another word for mean is average.
10	A	To find the median, find the middle number in the data set. One way to find the median is to order the numbers from least to greatest and cross out the numbers until the middle is reached. Here, the middle numbers is 1/3.
11	C	Mode is the value that appears most in a set of data. The list has a mode of 1 since that number appears most (3 times).
12	A	Remember: Mean is the average, median is the middle number when data is arranged numerically, and mode is the number that appears most often. Since mode is the value that appears most in a set of data, the original list has a mode of 14 because it is the number that appears most (2 times). Find the mean by adding the set of values and dividing by the amount of numbers in the set of data: (1) The sum of 6,8,1,5,11,14 and 14 = 59 (2) 59/7 = 8.4 To find the median, find the middle number in the data set. One way to find the median is to order the numbers from least to greatest and cross out the numbers until the middle is reached. Here, the middle number is 8.
13	A	Find the mean by adding the set of values and dividing by the amount of numbers in the set of data: (1) The sum of 0.3, 1.2, 2.5, 4.3 = 8.3 (2) 8.3/4 = 2.1 Therefore, the mean is 2.1.
14	C	To find the median, find the middle number in the data set. One way to find the median is to order the numbers from least to greatest and cross out the numbers until the middle is reached. Here, the middle numbers are 1.2 and 2.5. Add the numbers together and divide by 2. (1) 1.2 + 2.5 = 3.7 (2) 3.7 ÷ 2 = 1.85, which rounds to 1.9. Therefore, the median is 1.9
15	C	To find the mean absolute deviation, (1) find the mean, (2) find the difference between each data value and the mean, and (3) average the differences. Here, the mean is 3.4. The differences between the data values and mean are 1.4, 1.6, 3.6, 2.4 and 1.4. The average of the differences is 2.08.

Question No.	Answer	Detailed Explanation

Mean, Median, and Mode (7.SP.B.4)

Question No.	Answer	Detailed Explanation
1	B	Out of 16 scores, 13 of 16 are 4 or below. Since the majority of the scores are low, a person can infer that, overall, customers had a bad experience.
2	D	To answer this question, look at the results of the different dishes. Spaghetti: 7 Pizza: 13 Calzone: 9 Baked Ziti: 2 Since only 2 people said that baked ziti was their favorite food, it is the menu item the manager should remove from the menu.
3	A	The measures of central tendency are mean, which is the average of a set of data; median, which is the middle of a set of data, and mode, which is the value which appears most in a set of data.
4	A	To find the mean absolute deviation, (1) find the mean, (2) find the absolute value of the difference between each data value and the mean, and (3) average the differences. Here, the mean is 2.5. The difference between the data value and mean is 1.5, 0.5, 0.5, and 1.5. The average is 1.
5	A	Since mode is the value that appears most in a set of data, the original list has two modes of 2 and 13 because each value appears the same number of times. Find the mean by adding the set of values and dividing by the amount of numbers in the set of data: (1) 2 + 2 + 3 + 4 + 6 + 8 + 9 + 10 + 13 + 13 + 16 + 17 = 103 (2) 103/12 = 8.6 To find the median, find the middle number in the data set. One way to find the median is to order the numbers from least to greatest and cross out the numbers until the middle is reached. Here, the middle numbers are 8 and 9. Add the numbers together, and divide by 2. (1) 8 + 9 = 17 (2) 17 ÷ 2 = 8.5
6	D	Since mode is the value that appears most in a set of data, the original list has two modes of 11 and 23 since each value appears the same number of times. Find the mean by adding the set of values and dividing by the amount of numbers in the set of data: (1) 11 + 11 + 12 + 13 + 15 + 17 + 18 + 20 + 23 + 23 + 26 + 27 = 219 (2) 216/12 = 18. To find the median, find the middle number in the data set. One way to find the median is to order the numbers from least to greatest and cross out the numbers until the middle is reached. Here, the middle numbers are 17 and 18. Add the numbers together, and divide by 2. (1) 17 + 18 = 35 (2) 35 ÷ 2 = 17.

Question No.	Answer	Detailed Explanation
7	D	Since mode is the value that appears most in a set of data, the original list has two modes of 31 and 54 since each value appears the same number of times. Find the mean by adding the set of values and dividing by the amount of numbers in the set of data: (1) The sum of 31, 31, 32, 33, 35, 37, 38, 41, 54, 54, 57, 58 = 501 (2) 501/12 = 41.8 To find the median, find the middle number in the data set. One way to find the median is to order the numbers from least to greatest and cross out the numbers until the middle is reached. Here, the middle numbers are 37 and 38. Add the numbers together, and divide by 2. (1) 37 + 38 = 75 (2) 75 ÷ 2 = 37.5
8	A	Since mode is the value that appears most in a set of data, the original list has a mode of 64 since the number appears most (3 times). Find the mean by adding the set of values and dividing by the amount of numbers in the set of data: (1) The sum of 41, 41, 42, 43, 45, 47, 48, 51, 64, 64, 67, 64 = 617 (2) 617/12 = 51.4 To find the median, find the middle number in the data set. One way to find the median is to order the numbers from least to greatest and cross out the numbers until the middle is reached. Here, the middle numbers are 47 and 48. Add the numbers together, and divide by 2. (1) 47 + 48 = 95 (2) 95 ÷ 2 = 47.5
9	B	Since mode is the value that appears most in a set of data, the original list has a mode of 64 since the number appears most (3 times). Find the mean by adding the set of values and dividing by the amount of numbers in the set of data: (1) The sum of 51, 51, 52, 53, 55, 57, 58, 61, 54, 54, 57, 54 = 657 (2) 657/12 = 54.8 To find the median, find the middle number in the data set. One way to find the median is to order the numbers from least to greatest and cross out the numbers until the middle is reached. Here, the middle numbers are 54 and 54. Add the numbers together, and divide by 2. (1) 54 + 54 = 108 (2) 108 ÷ 2 = 54

Question No.	Answer	Detailed Explanation
10	A	Since mode is the value that appears most in a set of data, the original list has a mode of 61 since the number appears most (3 times). Find the mean by adding the set of values and dividing by the amount of numbers in the set of data: (1) The sum of 61, 61, 52, 53, 65, 67, 58, 61, 64, 64, 57, 54 = 717 (2) 717/12 = 59.8 To find the median, find the middle number in the data set. One way to find the median is to order the numbers from least to greatest and cross out the numbers until the middle is reached. Here, the middle numbers are 61 and 61. Add the numbers together, and divide by 2. (1) 61 + 61 = 122 (2) 122 ÷ 2 = 61
11	C	Since mode is the value that appears most in a set of data, the original list has two modes -- 71 and 74 -- since those numbers appear most (3 times). Find the mean by adding the set of values and dividing by the amount of numbers in the set of data: (1) The sum of 71, 71, 62, 63, 75, 77, 68, 71, 74, 74, 67, 74 = 847 (2) 847/12 = 70.6 To find the median, find the middle number in the data set. One way to find the median is to order the numbers from least to greatest and cross out the numbers until the middle is reached. Here, the middle numbers are 71 and 71. Add the numbers together, and divide by 2. (1) 71 + 71 = 142 (2) 142 ÷ 2 = 71
12	A	Since mode is the value that appears most in a set of data, the original list has two modes -- 71 and 74 -- since those numbers appear most (2 times). Find the mean by adding the set of values and dividing by the amount of numbers in the set of data: (1) The sum of 61, 71, 52, 63, 65, 77, 58, 71, 64, 74, 57, 74 = 787 (2) 787/12 = 65.6 To find the median, find the middle number in the data set. One way to find the median is to order the numbers from least to greatest and cross out the numbers until the middle is reached. Here, the middle numbers are 64 and 65. Add the numbers together, and divide by 2. (1) 64 + 65 = 129 (2) 129 ÷ 2 = 64.5

Question No.	Answer	Detailed Explanation
13	B	Since mode is the value that appears most in a set of data, the original list has a mode of 44 since that number appears most (3 times). Find the mean by adding the set of values and dividing by the amount of numbers in the set of data: (1) The sum of 21, 21, 22, 23, 35, 37, 38, 41, 44, 44, 47, 44 = 417 (2) 417/12 = 34.8 To find the median, find the middle number in the data set. One way to find the median is to order the numbers from least to greatest and cross out the numbers until the middle is reached. Here, the middle numbers are 37 and 38. Add the numbers together, and divide by 2. (1) 37 + 38 = 75 (2) 75 ÷ 2 = 37.5
14	C	Selecting 108 is the result of finding the sum of the six values in the table and dividing by 6. Selecting 110 results from choosing the mean per day the company needs to make a profit. A mean of 222 results from a calculation error when adding or subtracting the sum of the values and the needed number of visitors. The mean is the average of the values in a set of data. In order for Kelli's Ice Cream Shop to have a mean of 110 per day, multiply 110 times the number of days. 110 x 7 = 770. Kelli's Ice Cream Shop needs 770 visitors per week to make a profit. Add the number of visitors from the table: 63 + 77 + 121 + 96 + 137 + 154 = 648. Subtract the sum from 770. 770 - 648 = 122 Therefore, Kelli's needs 122 visitors on Saturday in order to have a mean of 110 visitors per day for the week and to turn a profit.
15	C	Since mode is the value that appears most in a set of data, the original list has no mode since each value appears the same number of times. If 90 is added to the list, the score would appear twice. Therefore, the mode would increase. The mean would neither increase nor decrease. The mean of the original list is 90, so adding a value of 90 would not change the mean. In the original list, the middle value of the numbers is 89. By adding 90 to the list, there will be two middle values, 89 and 90. The sum of the two middle values is 179. Divide the sum by 2. The quotient, 89.5, is the new median. Therefore, the median will increase by adding 90 to the list.

Question No.	Answer	Detailed Explanation
		Understanding Probability (7.SP.C.5)
1	D	Mary originally had 10 marbles in her bag. When she removed 3 marbles, she had 7 remaining marbles. Six of those marbles are yellow. Probability is: (chance of successful outcome)/(total number of outcomes). Plug in the numbers, and solve. 6 chances to pick yellow out of 7 outcomes = 6/7
2	A	There are four number 2 cards in a deck of 52. If one number 2 card is removed, that leaves three number 2 cards out of 51 cards. Probability is: (chance of successful outcome)/(total number of outcomes). Therefore, there is a 3 out of 51 chance that John will pick number 2 at random.
3	B	Maggie has a total of 46 coins, 6 of which are quarters. Probability is: (chance of successful outcome)/(total number of outcomes). Therefore, there is a 6 out of 46 chance that Maggie will pick a quarter. This simplifies to 3 out of 23.
4	A	Mark has a total of 71 bills, 8 of which are tens. Probability is: (chance of successful outcome)/(total number of outcomes). Therefore, there is an 8 out of 71 chance that Mark will pick a ten.
5	D	Moe has a total of 83 nuts, 33 of which are peanuts. Probability is: (chance of successful outcome)/(total number of outcomes). Therefore, there is a 33 out of 83 chance that Moe will pick a peanut.
6	B	Xavier has a total of 83 nuts, 8 of which are walnuts. If he removes the pecans, he will have 69 nuts remaining (83 - 14 = 69). Probability is: (chance of successful outcome)/(total number of outcomes). Therefore, there is an 8 out of 69 chance that Xavier will pick a walnut.
7	D	Tim had a total of 86 chocolates. If he removes 2 of each type of chocolate, he will have 22 cherry, 24 caramel, 18 fudge, and 14 candy remaining for a total of 78 chocolates. Probability is: (chance of successful outcome)/(total number of outcomes). Therefore, there is a 22 out of 78 chance that Tim will pick a cherry chocolate. This reduces to 11 out of 39.
8	A	Clarissa had a total of 86 chocolates. If she removes all of the taffy, she will have 70 chocolates remaining. Probability is: (chance of successful outcome)/(total number of outcomes). Therefore, there is a 20 out of 70 chance that Clarissa will pick a fudge chocolate. This reduces to 2 out of 7.

Question No.	Answer	Detailed Explanation
9	A	Karen had a total of 86 chocolates. If she removes 1/2 of the cherry and fudge chocolates, she will have 12 cherry, 26 caramel, 10 fudge, and 16 taffy remaining for a total of 64 chocolates. Probability is: (chance of successful outcome)/(total number of outcomes). Therefore, there is a 16 out of 64 chance that Karen will pick a taffy chocolate. This reduces to 1 out of 4.
10	C	Choosing 1/3 is the result of thinking Jessie will pick up 1 of 3 available peaches. Selecting 3/15 results from looking at the total number of peaches over the total numbers of fruit. A choice of 2/3 is the opposite of 1/3; it represents that Jessie will not pick up 2 of 3 peaches. [Since the probability of Jessie picking up a peach is 3/15, subtract that amount from 15/15 (the total number of fruits). 15/15 - 3/15 = 12/15, which in simplest form is 4/5. Therefore, the probability that he will not pick a peach is 4/5.]

Predicting Using Probability (7.SP.C.6)

1	C	Create a sample space by listing all the possible outcomes. For each coin flipped, it will land on heads or tails. Therefore, for two coins, there could be outcomes of (heads, heads), (heads, tails), (tails, heads), and (tails, tails).
2	A	When testing probability, larger samples (experiments) yield more accurate results. An experiment of 1,000 coin tosses could adequately test if a coin would land on its head 1/2 of the time (about 500 of 1,000).
3	D	Since there are three even numbers (2, 4, 6) and three odd numbers (1, 3, 5) on a six-sided die, the probability is 3 out of 6 for rolling either an even or an odd number. Therefore, it is equally likely to roll an even or odd number. There is also a 3 out of 6 chance of rolling a number from 1 to 3.
4	C	The theoretical probability (expected outcome) of flipping a coin is 1 out of 2 because it will either land on heads or tails (those are the two possible outcomes). 1 out of 2 is equivalent to 50%. Subtract the expected outcome from the actual outcome 75% (75 - 50 = 25) and add the percent sign: 25%

Question No.	Answer	Detailed Explanation
5	B	The theoretical probability (expected outcome) of rolling each die is 1 out of 4 because it will either land on 1, 2, 3 or 4 (those are the four possible outcomes). One out of 4 is equal to 25%. Therefore, the difference between the expected results and the actual results is 5% for each side.
6	A	The theoretical probability (expected outcome) of rolling each die is 1 out of 4 because it will either land on 1, 2, 3 or 4 (those are the four possible outcomes). So, the expected outcome would be 10 for each side. The actual outcome was 40(0.30) for side one and side four and 40(.20) for side two and side three, which equals 12 side one, 8 side two, 8 side three, and 12 side four.
7	A	If Juliana made 60 serves, multiply 60 by the decimal form of 80% to find her expected first serve number. $60 \times .80 = 48$ first serves expected. If she only made half of the serves, divide 60 by 2. $60 \div 2 = 30$ actual serves made.
8	A	Create a sample space by listing all the possible outcomes. Combining duplicate outcomes leaves 10 possible results: (1,1) (1,2) (1,3) (1,4) (2,2) (2,3) (2,4) (3,3) (3,4) (4,4)
9	D	There are 10 black cubes in the box and 25 total cubes. Therefore, Lea has a 10 out of 25 chance to pick a black cube. Multiply both 10 and 25 by 4 in order to find out her chances in percent form. $10 \times 4 = 40$ and $25 \times 4 = 100$. Since there is a 40 out of 100 chance to pick a black cube, Lea's chances of winning the game is 40%.
10	A	Since there are 16 fruit snacks out of a total of 46 total snacks, the probability of randomly picking a fruit snack is 16 out of 46. Divide 16 by 46, and then multiply by 100 in order to find the probability in percent form. $16 \div 46 = 0.3478 \times 100 = 34.78\%$. Rounding to the nearest tenth means a probability of 34.8% that a student will randomly select a fruit snack.

Question No.	Answer	Detailed Explanation

Using Probability Models (7.SP.C.7.A)

1	D	Remember: In a compound and event, you can multiply the probabilities of each event in order to arrive at a final solution. There is a 1/6 chance she will roll a 3 and a 1/6 chance Sara rolls a 5. This means there is a 1/6(1/6) = 1/36 chance she will roll both. Another way to approach this problem is to create the sample space containing all possible combinations. (B, Y): (1,1), (1,2), (1,3), (1,4), (1,5), (1,6) (2,1), (2,2), (2,3), (2,4), (2,5), (2,6) (3,1), (3,2), (3,3), (3,4), (3,5), (3,6) (4,1), (4,2), (4,3), (4,4), (4,5), (4,6) (5,1), (5,2), (5,3), (5,4), (5,5), (5,6) (6,1), (6,2), (6,3), (6,4), (6,5), (6,6) From this list it is clear that only one out of the 36 total possible matches the (3,5) described.
2	B	Probability is defined as chance that an event will occur (a number between 0 and 1). The closer a probability is to 1, the more likely it is to occur. Drawing a number line with marks 0.1 distance apart can help determine the likelihood of an event occurring. Since 0.91 is close to 1, it represents an event most likely to occur.
3	D	Probability is defined as the chance that an event will occur (a number from 0 to 1). Since 5/4 is a rational number greater than 1, it cannot represent a probability.
4	B	Theoretical probability is the (number of possible favorable outcomes)/(total number of outcomes). Each time Tom flips a coin, there is 1 possible favorable outcome (heads) out of 2 total possible outcomes. Therefore, the probability is 1/2, which is equivalent to 50%.
5	D	Since there are the same amount of coins in each amount, the theoretical probability (expected outcome) picking a coin is 1 out of 4 because Joe will either pick a penny, a nickel, a dime or a quarter. 1 out of 4 is equivalent to 25%.
6	B	For every roll of two 6-sided dice, there are 36 possible outcomes. Therefore, the probability of rolling a pair of fours is 1 out of 36.
7	C	For every roll of a 6-sided die and coin flip, there are 12 possible outcomes: {(1,h), (1,t), (2,h), (2,t), (3,h), (3,t), (4,h), (4,t), (5,h), (5,t), (6,h), (6,t)}. Therefore, the possibility of rolling a three and flipping tails is 1 out of 12.

Question No.	Answer	Detailed Explanation
8	B	For every roll of two 6-sided dice and two coin flips, there are 144 possible outcomes (36 x 4). Therefore, the probability of rolling two fives and flipping two heads is 1 out of 144.
9	A	The sample space of {BS, BM, BL, GS, GM, GL} is the correct answer. Since Sophia has a choice of 2 different pairs of shorts in 3 different combinations, the sample space should reflect choices of black or green shorts in sizes of small, medium, or large.
10	B	For each type of crust, there are 9 possible combinations. Therefore, the sample space {TPG, TPO, TPM, TIG, TIO, TIM, THG, THO, THM} represents the number of combinations if a customer chooses a thin crust pizza.

Probability models from observed frequencies (7.SP.C.7.B)

1	B	Because 5 of the 8 times it was flipped, the result was heads, the probability of heads is 5 out of 8.
2	A	If 20% of the rolls result in a 2, then 20% of 50 rolls would be (0.20)(50) = 10 times.
3	D	An even distribution of the odds and evens would be 25 each. 4 more odds would mean 27 odds and 23 evens.
4	A	4 out of 12 customers did NOT order pepperoni. This simplifies to 1 out of 3.
5	D	15 out of the 20 people ordered 1 box or more. This reduces down to 3 out of every 4 people.
6	A	Only 1 time out of 10 did he NOT pass the test, so this is a 10% probability.
7	C	7 out of 9 is a 77.777...% probability, which rounds to 78%.
8	B	An 80% success rate means that he makes 4 out of 5 and misses only 1 out of every 5 free throws.
9	A	The fact that 3 of the last 4 years they won the tournament suggests that this year's probability of winning is that same 3 out of 4.
10	B	4 out of the last 6 results were in the top three, so this simplifies to a 2 out of 3 probability.

Question No.	Answer	Detailed Explanation

Find the Probability of a Compound Event (7.SP.C.8.A)

1	B	For the t-shirt, polo shirt and sweater, determine how many outfits Jane can make counting the number of jeans and khakis and sneakers and sandals she can match with her shirts. Therefore, there are 12 combinations available for the outfits Jane can make.
2	C	For the t-shirt, polo shirt and sweater, determine how many outfits Jane can make, counting the number of jeans and khakis and sneakers and sandals she can match with her shirts. Therefore, there are 12 combinations available for the outfits Jane can make. For 12 outfits, there are 3 outfits with jeans and sneakers. Therefore, there is a 3 out of 12 probability that Jane will wear jeans and sneakers. 3 out of 12 is 3/12, which is 1/4 in simplest form.
3	C	For each name in first place, there are 24 ways the other names can be arranged. 24 × 5 = 120 ways.
4	D	For every roll of two 4-sided dice, there are 16 possible outcomes. Four of these outcomes would be favorable (1, 2), (2, 1), (1, 3), or (3, 1). So, the probability is 4 out of 16, or 1 out of 4.
5	B	For every roll of two 6-sided dice, there are 36 possible outcomes. Since there are 6 doubles combinations, the probability of rolling a double is 6 out of 36, or 1 out of 6.
6	A	There are 6 ways to arrange the players for the top two spots (combining duplicates): JDSC, JCDS, SCDJ, SJCD, CDSJ, SDJC.
7	C	For every three coin flips, there are 8 possible outcomes. 1 of those outcomes would contain all heads. So, the probability is 1 out of 8.
8	B	For every roll of a 6-sided die and a 4-sided die, there are 24 possible outcomes. Of these, 16 outcomes contain a 2 or a 3 (or both). This leaves 8 favorable outcomes. So the probability is 8 out of 24, or 1 out of 3.
9	B	For every roll of three 4-sided dice, there are 64 possible outcomes. 24 of these outcomes would contain one even and two odds. Therefore, the probability is 24 out of 64, or 3 out of 8.

Question No.	Answer	Detailed Explanation
10	A	To find the mean, add the numbers together, and divide by the amount of numbers in the data set. (1) 21 + 21 + 22 + 23 + 25 + 27 + 28 + 31 + 34 + 34 + 34 + 37 = 337 (2) 337 ÷ 12 = 28.1 To find the median, find the middle number in the data set. One way to find the median is to order the numbers from least to greatest and cross out the numbers until the middle is reached. Here, the middle numbers are 27 and 28. Add the numbers together and divide by 2. (1) 28 + 27 = 55 (2) 55 ÷ 2 = 27.5 To find the mode, find the number that appears most in the data set. Looking at the numbers shows that the number 34 appears 3 times; therefore 34 is the mode.

Represent sample spaces (7.SP.C.8.B)

Question No.	Answer	Detailed Explanation
1	A	All possible combinations of heads (H) and tails (T) must be named; there are four such combinations.
2	A	All possible combinations of the numbers 1 through 6 must be given in the table. There are 36 different possible combinations.
3	D	The blue stone might be either the first or second of the stones. The other stone could be either red or green.
4	D	If three of the checkers are red, then one is black. The black could be drawn out as the first, second, third, or fourth of the checkers, so there are four ways this could take place.
5	B	The number that is chosen both times could be any of the 10 numbers.
6	A	There are 5 even numbers to be followed by any of 4 prime numbers. These can be combined 20 different ways.
7	C	The sample space for this outcome would be: (1, 6), (2, 5), (3, 4), (4, 3), (5, 2), and (6, 1). There are six possible outcomes.
8	D	The first card drawn could be any of the cards other than the 3. There are 19 such possibilities. The second card could be anything but the 3 and whatever card happened to be drawn first, which gives 18 possibilities. The third card must be the 3. It doesn't matter what the cards are after that. There are 342 ways to combine these possibilities.
9	B	The first arrow could be in any of the four rings. The second arrow could also be in any of the four rings. There are 16 ways to combine these possibilities.

Question No.	Answer	Detailed Explanation
10	A	The first player could be in any of 5 positions; the second in any of the remaining 4 positions; the third in any of the remaining 3 positions; the fourth in any of the remaining 2 positions; the fifth must be placed in the only remaining position. The total number of ways to combine these options can be found by multiplying the number of possibilities for each placement: $(5)(4)(3)(2)(1) = 120$ possible outcomes.

Simulate compound events to estimate probability (7.SP.C.8.C)

1	A	The probability that the first will be male is 80%. Multiply by 80% again to find the probability that the second is also male.
2	A	The slips of paper should be in the same ratio as the population, which can be done by 2 adult slips and 3 child slips.
3	C	The probability that the second customer will choose the same bread as the first is 1 in 6. The probability that he will choose the same meat is 1 in 5. Multiply these together to find the probability of choosing both.
4	B	The probability of choosing different shorts is 5 out of 6. The probability of choosing different shirts is 3 out of 4. Multiplying these together to find the probability of both gives us 5 out of 8.
5	A	The probability of choosing the same pants is 1 in 3. The probability of choosing the same shoes is 1 in 5. Multiply these together to get the probability of the same combination of both, which is 1 in 15.
6	C	You must get 5 consecutive flips without rain to represent five days without rain. You must also test this numerous times and record the percentage of the time that no rain is the result.
7	A	The probability of it starting well one time is four out of five. You must multiply this together three times, though, to find the probability of it starting well all three times that you have to mow. The result is about 51%.

Question No.	Answer	Detailed Explanation
8	C	He has a 1 in 2 chance of being on time one day. Multiply this together three times to find the likelihood that he will be on time all three remaining days. The probability is 1 in 8.
9	B	The first friend could be born on any of the seven days of the week. The probability that the second friend was born on the same day as the first is 1 in 7. The probability that the third was born on that same day is also 1 in 7. Multiply these probabilities together to find the probability that they were all born on the same day. 1 in 49 is the answer.
10	D	The probability that they chose the same model is 1 in 4. The probability that they chose the same color is 1 in 6. Multiply these together to find the probability that they chose the same of both.

Notes

SBAC FAQ

What will SBAC Math Assessments Look Like?

In many ways, the SBAC assessments will be unlike anything many students have ever seen. The tests will be conducted online, requiring students complete tasks to assess a deeper understanding of the CCSS. The students will be assessed once 75% of the year has been completed in two different assessments - a Computer Adaptive Testing (CAT) and a Performance Task (PT).

The time for each Math portion is described below:

Estimated Time on Task in Minutes		
Grade	CAT	PT
3	90	60
4	90	60
5	90	60
6	120	60
7	120	60
8	120	60

Bacause the assessment is online, the test will consist of a combination of new types of questions:

1. Drag and Drop
2. Drop Down
3. Essay Response
4. Extended Constructed Response
5. Hot Text Select and Drag
6. Hot Text Selective Highlight
7. Matching Table In-line
8. Matching Table Single Reponse
9. Multiple Choice – Single Correct Response, radial buttons
10. Multiple Choice – Multiple Response, checkboxes
11. Numeric Response
12. Short Text
13. Table Fill-in

What is this SBAC Test Practice Book?

Inside this book, you will find practice sections aligned to each CCSS. Students will have the ability to review questions on each standard, one section at a time, in the order presented, or they can choose to study the sections where they need the most practice.

In addition to the practice sections, you will have access to two full-length CAT and PT practice tests online. Completing these tests will help students master the different areas that are included in newly aligned SBAC tests and practice test taking skills. The results will help the students and educators get insights into students' strengths and weaknesses in specific content areas. These insights could be used to help students strengthen their skills in difficult topics and to improve speed and accuracy while taking the test.

Because the SBAC assessment includes newly created, technology-enhanced questions, it is necessary for students to be able to regularly practice these questions. The Lumos online StepUp program includes thirteen technology enhanced questions that mimic the types students will see during the assessments. These include:

1. Drag and Drop

This style of question requires the student to move the correct answer from the answer choices into a box. Usually, this box is below the answer choices. If there is more than one box, click on the answer and drag it to the correct box. Hold the answer until it reaches the box and then release the mouse. Often, with multiple boxes, the student may need to continue holding and dragging down as the page scrolls to the correct box.

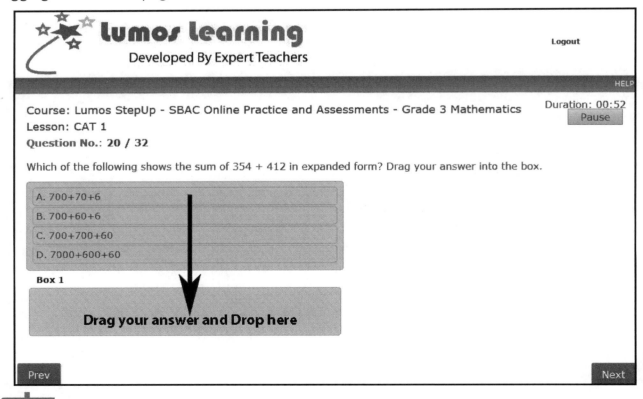

2. Drop Down

This style of question requires students to click on the drop down arrow and pull down the menu so they can select the correct answer.

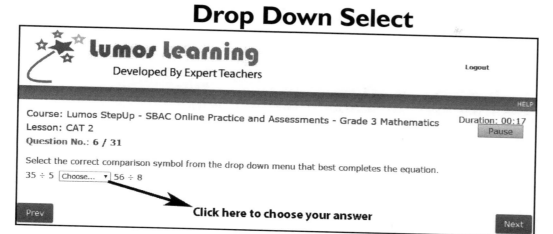

Drop Down Select

Course: Lumos StepUp - SBAC Online Practice and Assessments - Grade 3 Mathematics
Lesson: CAT 2
Question No.: 6 / 31

Select the correct comparison symbol from the drop down menu that best completes the equation.

$35 \div 5$ [Choose... ▼] $56 \div 8$

Click here to choose your answer

3. Essay Response

Essay response questions may be familiar to many students as they have been completing these types of responses for many years. The technology enhanced portion of this question though has the students typing their essay response into the box. As they type, the box will expand. Common word processing tools are available.

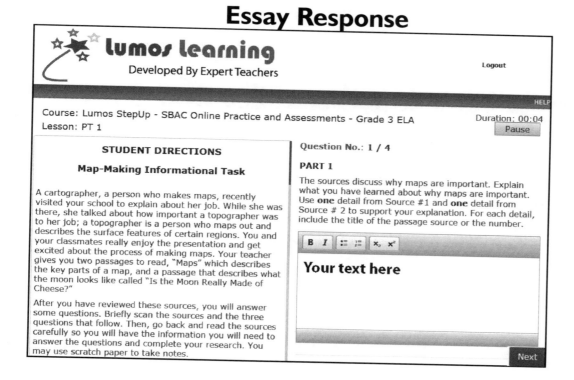

Essay Response

Course: Lumos StepUp - SBAC Online Practice and Assessments - Grade 3 ELA
Lesson: PT 1

Question No.: 1 / 4

STUDENT DIRECTIONS

Map-Making Informational Task

A cartographer, a person who makes maps, recently visited your school to explain about her job. While she was there, she talked about how important a topographer was to her job; a topographer is a person who maps out and describes the surface features of certain regions. You and your classmates really enjoy the presentation and get excited about the process of making maps. Your teacher gives you two passages to read, "Maps" which describes the key parts of a map, and a passage that describes what the moon looks like called "Is the Moon Really Made of Cheese?"

After you have reviewed these sources, you will answer some questions. Briefly scan the sources and the three questions that follow. Then, go back and read the sources carefully so you will have the information you will need to answer the questions and complete your research. You may use scratch paper to take notes.

PART 1

The sources discuss why maps are important. Explain what you have learned about why maps are important. Use **one** detail from Source #1 and **one** detail from Source # 2 to support your explanation. For each detail, include the title of the passage source or the number.

Your text here

4.   Extended Constructed Response

Similar to Essay Response, the ECR allows students to write their responses to the question. These answers are not as long as essay responses but they are usually longer than just one or two words.

Extended Constructed Response

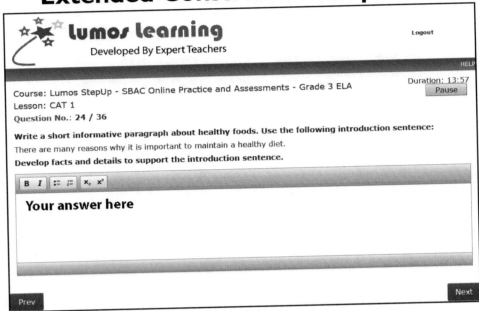

5.   Hot Text Select and Drag

With this style of questions, students must rearrange the text into the correct order. They might be asked to place events in a timeline, place an introductory sentence in the correct place, order information in the correct sequence, or any variety of tasks. Like a Drag and Drop, students can click on the highlighted sentence and move it where they would like it to be.

Hot Text - Select and Drag

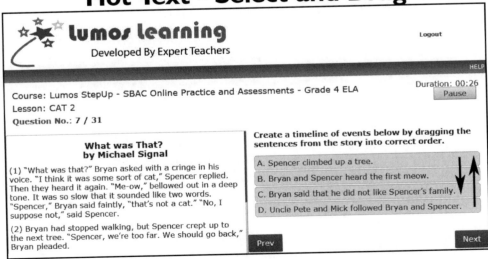

6.   Hot Text Selective Highlight

Hot Text – Selective Highlight asks students to select certain words or phrases from the paragraph for their answer. Often, this type of question will be used when students are asked for supporting information to defend an answer.

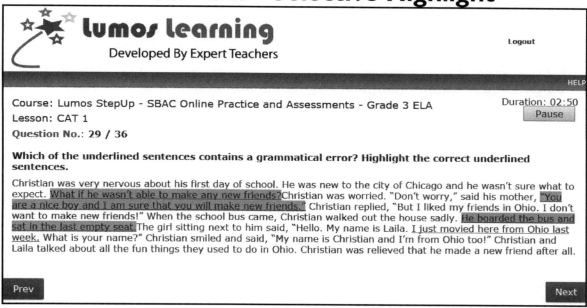

7.   Matching Table In-Line

This technology enhanced question requires students to process data spread across a table and mark check boxes. They may have one, or more than one, box selected in the table.

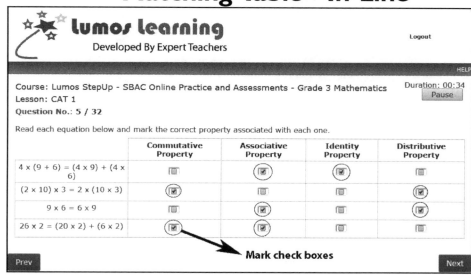

8.   Matching Table Single Response

Similar to the previous, Matching Table, students will make a selection in the table. With this type of question they will make a single choice such as yes/no or true/false.

Matching Table - Single Response

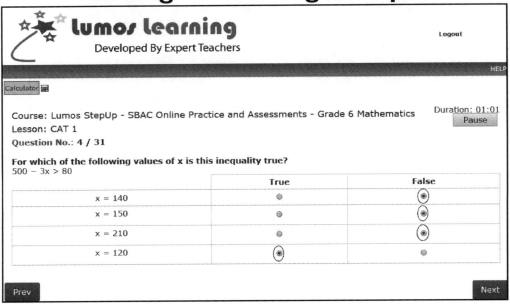

9.   Multiple Choice, Single Answer

This style of question is most similar to what students might recognize. It is a standard multiple choice with one answer.

Multiple Choice - Single Response

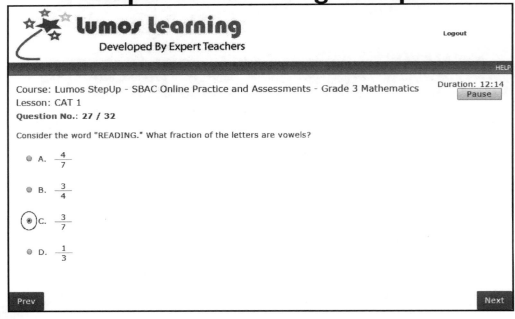

10. Multiple Choice, Multiple Answer

Similar to the previous style, this question asks students to make a selection from options below. However, with the MCMA question, students will need to choose more than one answer. Careful reading of the question is required as it may offer guidance to the number of answer that need to be selected.

Multiple Choice - Multiple Select

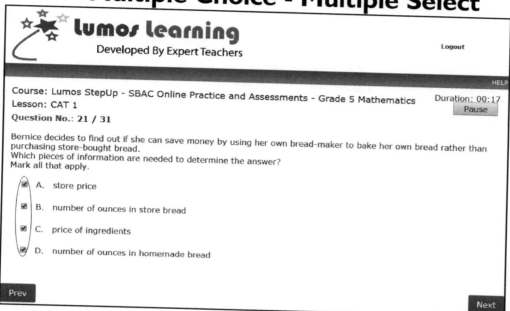

11. Numeric Response

Numeric response questions have a small box where students can type in their solution to a problem. They might use numbers, words, or any combination of both. The question will typically offer guidance to what will go in the box.

Numeric Response

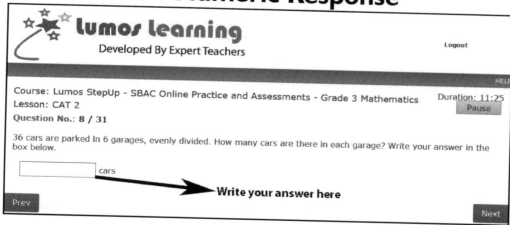

12.   Short Constructed Response

Similar to an Extended Constructed Response, this question asks students to write short responses. Typically, students will use this box to explain how they arrive at a solution or why a response may be correct or incorrect.

Short Constructed Response

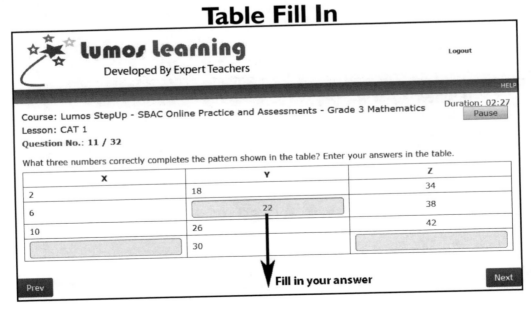

13.   Table Fill In

This style of question requires students to fill in their answers.

Table Fill In

© Lumos Information Services 2015 | LumosLearning.com

How is this Lumos tedBook aligned to SBAC Guidelines?

The SBAC practice tests offered online at Lumos Learning have been created to accurately reflect the depth and rigor of SBAC. Students will still be exposed to the technology enhanced questions so they become familiar with the wording and how to think through these types of tasks.

This edition of the practice test book was created in the FALL 2015 and aligned to the most current SBAC standards released to date. Some changes will occur as SBAC continues to release new information in the spring of 2016 and beyond.

Lumos StepUp® Mobile App FAQ For Students

What is the Lumos StepUp® App?

It is a FREE application you can download onto your Android smart phones, tablets, iPhones, and iPads.

What are the Benefits of the StepUp® App?

This mobile application gives convenient access to Practice Tests, Common Core State Standards, Online Workbooks, and learning resources through your smart phone and tablet computers.

- Eleven Technology enhanced question types in both MATH and ELA
- Sample questions for Arithmetic drills
- Standard specific sample questions
- Instant access to the Common Core State Standards
- Jokes and cartoons to make learning fun!

Do I Need the StepUp® App to Access Online Workbooks?

No, you can access Lumos StepUp® Online Workbooks through a personal computer. The StepUp® app simply enhances your learning experience and allows you to conveniently access StepUp® Online Workbooks and additional resources through your smart phone or tablet.

How can I Download the App?

Visit **lumoslearning.com/a/stepup-app** using your smart phone or tablet and follow the instructions to download the app.

QR Code
for Smart Phone
Or Tablet Users

Lumos SchoolUp™ Mobile App FAQ
For Parents and Teachers

What is the Lumos SchoolUp™ App?

It is a FREE App that helps parents and teachers get a wide range of useful information about their school. It can be downloaded onto smartphones and tablets from popular App Stores.

What are the Benefits of the Lumos SchoolUp™ App?

It provides convenient access to

- School "Stickies". A Sticky could be information about an upcoming test, homework, extra curricular activities and other school events. Parents and educators can easily create their own sticky and share with the school community.
- Common Core State Standards.
- Educational blogs.
- StepUp™ student activity reports.

How can I Download the App?

Visit **lumoslearning.com/a/schoolup-app** using your smartphone or tablet and follow the instructions provided to download the App. Alternatively, scan the QR Code provided below using your smartphone or tablet computer.

**QR Code
for Smart Phone
Or Tablet Users**

The Lumos Learning Teacher Portal gives teachers insights into their students' work and access to useful resources. The personalized teacher dashboard includes six different tabs.

1. Student Reports

The 'Student Report' tab is the heart of the Teacher Portal. It is where teachers can create their student accounts, glean the most information about their list of students, class performance, individual student performance, and explore their list of subscribed content.

Progress Summary

This tab allows teachers to follow the progress of each of their students. From here, teachers can see overall progress of each student. They can then click on a specific student's name and see the individual details of progress. Additionally, teachers, can click on questions to be graded (essays and constructed responses for example).

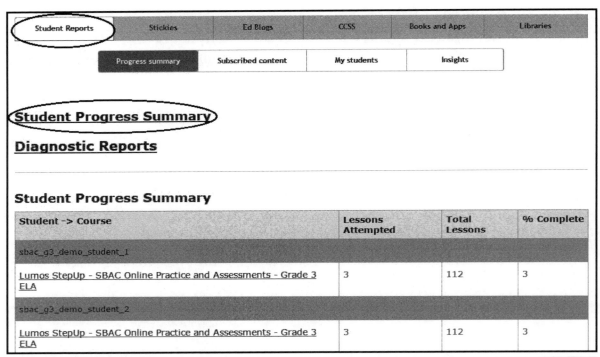

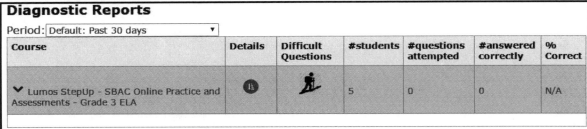

Subscribed Content

The subscribed content tab allows teachers to see all content to which they have subscribed. It is also from this tab that teachers can assign specific work to their students. A teacher can assign an individual lesson to the student through the 'View Worksheets' link; from there, the teacher can select from the list of lessons available. Students will then receive an alert about the assigned lesson.

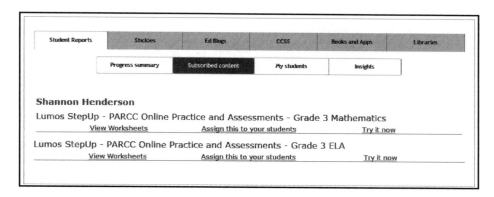

My Students

This tab allows teachers to see all login information for their assigned students. They can change passwords and create student accounts in this tab.

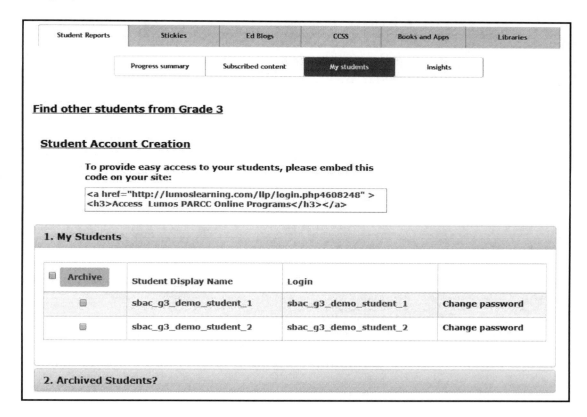

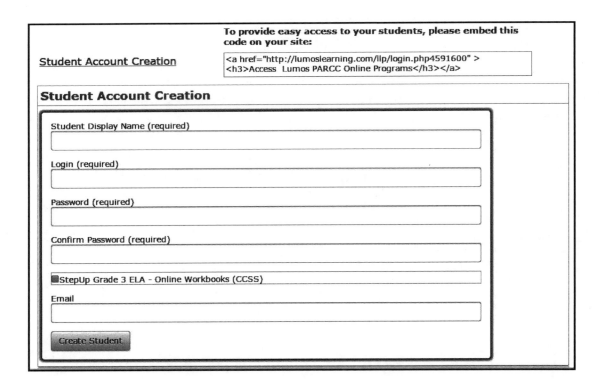

Insights

The insights tab is one of the most powerful parts of the teacher portal. It allows teachers to gain a deeper understanding of how their students are progressing. Individualized reports can be generated for specific date ranges. Student performance data can be categorized into partial, proficient, and advanced. Additionally, teachers can customize what guidelines they would like to stand for not meeting the standard (low bar), meeting the standard, and excelling above the standard. This report can be used to drive instruction and practice and identify what topics students might need additional assistance in to master the standard.

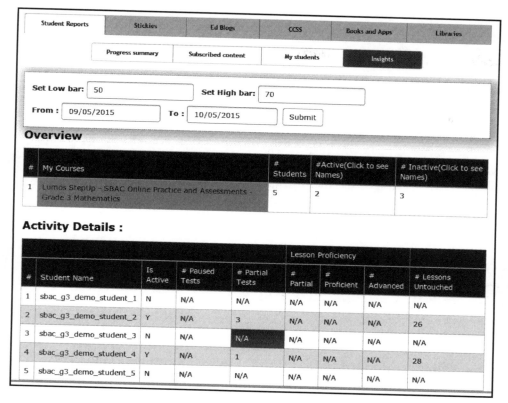

2. Stickies

Stickies are an exciting new way to share any type of school related information with parents and students.

- Need to share your school supply list?
- Have a great resource to exchange with others?
- Want to ensure parents can see a copy of the homework?
- Want to recommend a mobile app?
- Want to suggest a book?

Create a Stickie!

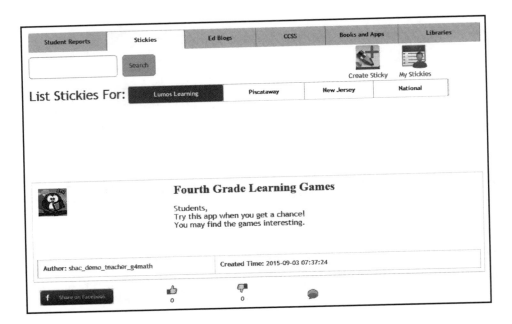

3. Ed Blogs

Lumos Learning teachers consistently monitor current educational trends and topics. Exploring the EdBlogs tab allows teachers to follow the blogs and stay current on important educational topics.

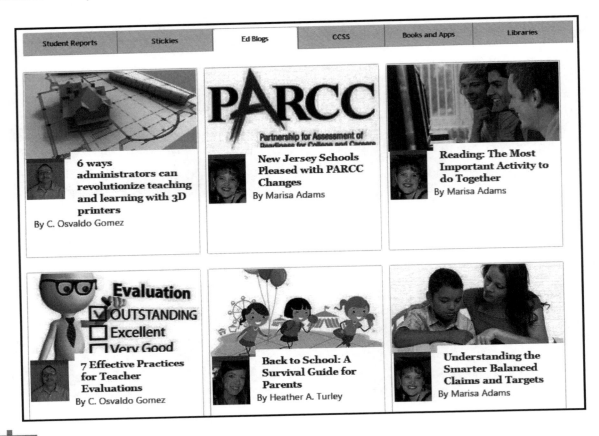

4. CCSS

With this tab, teachers are able to access the Common Core State Standards in one easy location. This eliminates the need to search in a variety of browsers to locate key information.

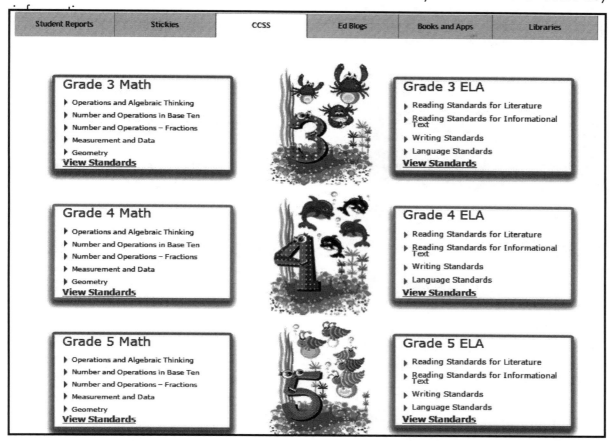

5. Books and Apps

This tab allows teachers to search for relevant educational books and apps. Teachers can easily recommend useful apps and books to their students by creating stickies.

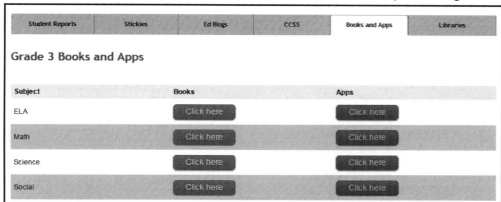

6. Libraries

Teachers can search their local libraries to look for educational books and other resources available in their area.

Student Reports	Stickies	Ed Blogs	CCSS	Books and Apps	Libraries

Libraries in Piscataway

Library Name	Address	Phone number
WESTERGARD LIBRARY	20 MURRAY AVE. PISCATAWAY Zip:8854	2017521166
PISCATAWAY PUBLIC LIBRARY	500 HOES LANE PISCATAWAY Zip:8854	2014631633

Lumos StepUp™ is an educational app that helps students learn and master grade-level skills in Math and English Language Arts.

The list of features includes:

- Learn Anywhere, Anytime!

- Grades 3-8 Mathematics and English Language Arts

- Get instant access to the Common Core State Standards

- One full-length sample practice test in all Grades and Subjects

- Full-length Practice Tests, Partial Tests and Standards-based Tests

- 2 Test Modes: Normal mode and Learning mode

- Learning Mode gives the user a step-by-step explanation if the answer is wrong

- Access to Online Workbooks

- Provides ability to directly scan QR Codes

- And it's completely FREE!

http://lumoslearning.com/a/stepup-app

Grade 7

Developed By Expert Teachers

SBAC Test Prep
ENGLISH
LANGUAGE ARTS LITERACY

Smarter Balanced Study Guide

ONLINE

2 Performance Tasks (PT)

2 Computer Adaptive Tests (CAT)

40+ SKILLS

Available

- At Leading book stores

- Online www.LumosLearning.com

9R00114

Made in the USA
San Bernardino, CA
06 April 2016